QUANTUM MECHANICS
Principles and Formalism

QUANTUM MECHANICS
Principles and Formalism

ROY MCWEENY
Professor Emeritus
University of Pisa, Italy

DOVER PUBLICATIONS, INC.
Mineola, New York

Bibliographical Note

This Dover edition, first published in 2003, is a republication of the work first published in 1972 by Pergamon Press, Ltd., Oxford, as Volume 1 in Topic 2 (Classical and Quantum Mechanics) of The International Encyclopedia of Physical Chemistry and Chemical Physics.

Library of Congress Cataloging-in-Publication Data

McWeeny, R.
    Quantum mechanics : principles and formalism / Roy McWeeny.
      p. cm.
    "This Dover edition ... is a republication of the work first published in 1972 by Pergamon Press, Ltd., Oxford, as volume 1 in topic 2 (Classical and quantum mechanics) of the International encyclopedia of physical chemistry and chemical physics"—T.p. verso.
    Includes bibliographical references and index.
    ISBN 0-486-42829-X (pbk.)
    1. Quantum theory. I. Title.

QC174.12 .M39 2003
530.12—dc21

                    2002072872

Manufactured in the United States of America
Dover Publications, Inc., 31 East 2nd Street, Mineola, N.Y. 11501

CONTENTS

PREFACE

Note: The following Preface reflects the fact that this book, now reprinted separately, was first published as part of The International Encyclopedia of Physical Chemistry and Chemical Physics.

No MAJOR work covering the general area of Chemical Physics would be complete without a substantial section on quantum mechanics. Quantum mechanics provides not only the physical principles and mathematical methods used in all theoretical work on the electronic structure and properties of matter; it provides even the concepts and vocabulary of nearly all branches of theoretical chemistry. It is therefore of some importance to indicate the aims and scope of this section of the Encyclopedia, distinguishing clearly between quantum mechanics as a discipline and quantum mechanics applied to chemical problems.

Quantum mechanics is here developed as a fundamental discipline. Although applications are frequently considered, they are introduced only to illustrate principles or give point to the development of the subject in various directions. More detailed applications of quantum mechanics—to atomic structure, molecular binding, molecular properties, solid state theory, and so on—are found in other sections of the Encyclopedia.

Each volume is severely restricted in length, covering a well-defined area at a fundamental level and exhibiting a considerable degree of independence so as to be useful as a textbook in its own right. Another prerequisite is that the exposition shall not be so severe as to overtax the resources of good graduate students in physics or chemistry, though clearly topics of intrinsic mathematical difficulty will be heavy reading for students whose mathematical skills are rudimentary.

With these limitations in mind it seemed necessary to plan for at least four short volumes on (non-relativistic) quantum mechanics. The first would confine itself to an exposition of the basic principles and formalism, illustrated mainly by reference to the simplest possible one-particle systems. The second would develop the main techniques—such as perturbation theory, and group theory—in the context of basic applications, again as far as possible with reference to a single particle moving in some given field: it would give point to the theory by introducing the "independent particle model", in which each electron moves in an "effective field" provided by the others, and in this way would serve to indicate the applicability of quantum mechanics to atoms, molecules and crystals. The third would deal quite generally with *many*-electron systems, introducing specifically "many-body" topics such as permutation symmetry, vector coupling, electron correlation, and would include general theory of electronic properties. The fourth would return to the basic principles and formalism, particularly those referring to time-dependent effects, and develop the more elementary parts of the theory of scattering and collisions. Each volume would thus be relatively compact and suitable for use either by itself or in conjunction with the others, according to the particular interests or background of the reader. Further volumes would no doubt be added as the pressure of new developments and applications, for example in more advanced collision theory, made them necessary.

The present volume is accordingly concerned with the principles and formalism of quantum mechanics. It is not strictly an introduction to the subject, since all graduate students in physics and chemistry can nowadays confidently be expected to have some acquaintance with quantum mechanics at its most elementary level. Chapters 1 and 2 merely review the origins of Schrödinger's equations and the nature of the solutions in certain simple and well-known cases. Matters which are extraneous to the subject (e.g. standard methods of separating and solving partial differential equations) are relegated to Appendices so as not to interrupt the development. The ideas associated with vector spaces, however, and more generally with Hilbert space, are so much a part of the whole fabric of quantum mechanics that they have been developed in Chapter 3 as an essential part of the text. This chapter provides the mathematical language with which it is possible to formulate the main principles and their immediate consequences in Chapter 4. Special attention has been given to the more difficult topics, such as spin, which are frequently skipped, glossed over or presented in an abstruse and unpalatable form. This chapter leads naturally to the final generalizations (Chapter 5) in which various alternative "languages" or representations are discussed—each with its own advantages in particular applications—and the Dirac transformation theory is developed and explained.

Any author of yet another book on quantum mechanics must expect criticism. A mathematical physicist may complain that the delta function has been introduced too casually, or that the Lagrangian formalism has not been given due prominence, or that continuing controversies about the measurement process have been largely ignored. Many a chemist, on the other hand, may object to so many pages on vector spaces and mathematical techniques. But the book is addressed specifically to neither of these readers. In Volume 1 I have tried to present the main principles of the theory with special regard to the needs of chemical physicists, recognizing that present work in such diverse fields as electron spin resonance, molecular beam experiments, and Compton scattering brings with it the need for much sophisticated analysis, and a familiarity with the Heisenberg representation, momentum space, and all the trappings of the general theory. If this and succeeding volumes can bring the theory and its current applications in chemical physics within the reach of a good graduate student they will fulfil their main object.

Finally, I wish to thank Professor K. Ohno (University of Hokkaido) for his help in the early planning of the first two volumes, which were started several years ago under joint authorship. Although distance and the pressure of other commitments prevented the full fruition of this collaboration, his contributions, and particularly his critical and constructive comments on most of the present volume in a semi-final form, are gratefully acknowledged. My thanks are also due to Mrs. S. P. Rogers for her careful and accurate production of the typescript, and to Pergamon Press for their co-operation at all times.

R. McW.

CHAPTER 1

PHYSICAL BASIS OF QUANTUM THEORY

1.1. Particles and waves

There is little doubt that the most concise and elegant exposition of the principles of quantum mechanics consists of a set of basic propositions, from which the whole theory may be derived without further appeal to experiment. The experimental basis of the subject is in this way absorbed into a set of postulates which, although by no means self-evident, lead to a network of conclusions which may be tested and verified. The postulates consequently provide a very succinct expression of the results of a wide range of observations. From these postulates, it is possible in principle, though often difficult in practice, to follow the ramifications of the theory into many branches of physics and chemistry.

In spite of the many attractions of the axiomatic approach, to which we return in Chapter 4, it is useful first to recall some of the basic observations concerning wave and particle behaviour of light and electrons. These led to the generalizations on which the more formal theory is based. To this end, some familiarity with the wave-particle "dualism" will be assumed. We recall two of its main features:

(A) There is evidence (e.g. from the photo-electric effect and the Compton effect) that radiation exhibits particle properties. It appears to be transmitted in localized "packets" with energy E and momentum p related to frequency v in the following way:

$$E = hv, \tag{1.1}$$

$$p = \frac{hv}{c}. \tag{1.2}$$

Here c is the velocity of light, h is Planck's constant;

$$c = 2 \cdot 997925 \times 10^8 \text{ m s}^{-1},$$

$$h = 6 \cdot 6256 \times 10^{-34} \text{ J s}.$$

A "light particle" with energy given by (1.1) and momentum by (1.2) is called a *photon*.

(B) There is evidence (e.g. from electron-diffraction experiments) that material particles exhibit wave properties. The relative frequency with which particles are found in a given region of space (measured, for example, by the intensity of darkening of a photographic plate on which a beam of particles falls) is found to be correctly predicted as the squared amplitude of a wave-like disturbance, propagated according to laws formally similar to those of physical optics. For particles travelling in a beam, with a constant velocity, the associated wave is plane and has its normal in the direction of motion. It was suggested by de Broglie, on the basis of relativistic considerations, that the wave length should be related to the particle momentum $p = mv$ by

$$\lambda = \frac{h}{p} \qquad (1.3)$$

which agrees exactly with (1.2) since $\lambda = c/v$, and this conjecture was subsequently verified experimentally.

Wave mechanics, the particular formulation of quantum mechanics due to Schrödinger, arose in the attempt to reconcile the apparent coexistence of wave-like and particle-like properties in both material particles and photons. Here we indicate the argument, in a rudimentary form, by considering a harmonic wave travelling in the positive x direction:

$$\psi(x, t) = A \exp \{\pm 2\pi i k(x - ut)\}, \qquad (1.4)$$

where ψ measures the magnitude of the disturbance at point x and time t and u is the velocity of propagation.

We remember the interpretation of k. If x increases by $1/k$, the values of ψ and its derivatives are unchanged: the disturbance is therefore *periodic in space*, at any given time, with period $\lambda = 1/k$. λ is the *wave length* and k is the *wave number*. Also if t increases by $1/(ku)$, ψ and its derivatives are again unchanged: the disturbance is therefore *periodic in time* at any given point in space, with period $T = 1/(ku)$. T is the *period* of a complete oscillation, its reciprocal $v = ku$ being the *frequency* of oscillation. The definitions are thus

$$k = \text{wave number}, \qquad\qquad \lambda = 1/k = \text{wave length},$$

$$v = ku = \text{frequency}, \qquad\qquad T = 1/v = \text{period}.$$

Also, the two functions (1.4) (either choice of sign) clearly oscillate only between maximum and minimum values $\pm A$; A is the *amplitude*. A disturbance which is everywhere real can of course be regarded as the real part of either function, or as the sum of the two since $e^{i\theta} = \cos \theta + i \sin \theta$.

Wave packets

According to Fourier's theorem, *any* disturbance, travelling to the right with velocity u, can be represented by combining waves such as (1.4) with variable k values and suitably chosen amplitudes:

$$\psi(x, t) = \int_{-\infty}^{+\infty} A(k) \exp\{2\pi i k(x-ut)\}dk \qquad (1.5)$$

which is the limit of a sum of terms

$$A_1 \exp\{2\pi i k_1(x-ut)\} + A_2 \exp\{2\pi i k_2(x-ut)\} + \ldots$$

as the values of k_1, k_2, ... get closer together until they cover the whole range $(-\infty, +\infty)$,‡ the amplitude then becoming a continuous function of k. An arbitrary disturbance which is everywhere real can always be represented in this form by suitably choosing the amplitudes; thus by taking just two terms with $k_1 = -k_2 = k$ and $A_1 = A_2$ we obtain a travelling wave with $\psi \propto \cos 2\pi k(x-ut)$. Equivalently, we may simply take the real part of an expression of the form (1.5).

We now try to obtain a *localized* disturbance—suitable for associating with a moving particle—by choosing the amplitude factor $A(k)$ so that the wave trains admitted interfere constructively at one point but destructively at all others. A *wave packet* of this kind can in fact be constructed by using a very narrow range of wave numbers and may therefore be virtually homogeneous in frequency. If radiation (or matter) were propagated in wave packets, each packet could thus correspond to a sharply specified frequency and yet be essentially localized like a particle. To illustrate this possibility we consider a "Gaussian" wave packet in which

$$A(k) = a \exp\{-\sigma(k-k_0)^2\}. \qquad (1.6)$$

When σ is large this will describe a superposition of wave trains with wave numbers differing inappreciably from k_0, the amplitudes falling off rapidly according to the Gaussian law. In this case,

$$\psi(x, t) = \int_{-\infty}^{\infty} a \exp\{-\sigma(k-k_0)^2 + 2\pi i k(x-ut)\}dk. \qquad (1.7)$$

From the well-known result

$$\int_{-\infty}^{\infty} \exp\{-px^2 - qx\}dx = \sqrt{\frac{\pi}{p}} \exp\left[\frac{q^2}{4p}\right] \qquad (1.8)$$

‡Actually $|k|$ is then the wave number: by including the negative values we simply admit both signs of k in (1.4). The notation (a, b) in general denotes an *interval*, i.e. the range of values taken by a variable.

(which is valid when p and q are complex, provided the real part of p is positive), it follows easily that

$$\psi(x, t) = a\sqrt{\frac{\pi}{\sigma}} \exp\left(-\frac{\pi^2(x-ut)^2}{\sigma}\right) \exp\{2\pi ik_0(x-ut)\} \qquad (1.9)$$

At $t = 0$, this describes an oscillation of wave length $\lambda_0 = 1/k_0$, whose amplitude $a\sqrt{\pi/\sigma}\,\exp(-\pi^2x^2/\sigma)$ is rapidly damped about the point $x = 0$ (Fig. 1.1a).

In the case of a photon, $u = c$ (the velocity of light) is constant in free space. The packet is then propagated without dispersion (i.e. change of shape) for $\psi(x, t)$ at time t has exactly the same form as $\psi(x, 0)$ except that x is replaced by $x - ct$, i.e. the pattern is shifted to the right through a distance ct (Fig. 1.1b). The frequency ck_0, from

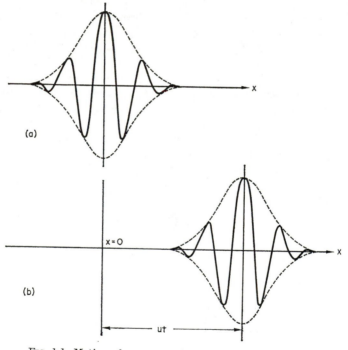

FIG. 1.1. Motion of a wave packet: (a) initial form ($t = 0$), (b) form at time t.

which the waves in the packet deviate inappreciably, is presumably related to the photon energy by the Planck law (1.1), and the photon, identified with the wave packet, travels with velocity c.

In the case of a material particle, it might seem natural to relate the frequency of an associated wave train to the particle energy in the same way as for a photon, namely by the Planck law. But the fact that λ is known to depend on the particle momentum, according to de Broglie's relationship (1.3), then implies that the velocity of propagation, $u = \lambda v$, is no longer a constant but depends on the dynamical situation. We tentatively accept this implication and combine (1.1) and (1.3) to obtain (since $p = mv$ and $E = \frac{1}{2}mv^2 + V$)

$$u = \lambda v = \frac{h}{p}\frac{E}{h} = \frac{E}{\sqrt{\{2m(E-V)\}}} \qquad (1.10)$$

where E is the total energy of the particle, and V its potential energy in the field in which it moves. The velocity of propagation then depends on frequency (through E) and, when the potential is non-uniform, on the particle position (through V). In this case *dispersion* occurs as the wave packet is propagated, and it is necessary to distinguish *two* velocities: u is now referred to as the *phase velocity*, but no longer coincides with the velocity of the packet itself, i.e. the velocity of the point at which the component waves reinforce. The wave packet moves with the *group velocity*, v_g, which may differ considerably from u.

We now show that by accepting (1.10) we ensure that v_g exactly coincides with the particle velocity v. We again consider the Gaussian packet (1.7) but now admit dispersion by assuming a non-linear relationship between k and v, as would follow from (1.10). Since the integrand in (1.7) is small unless k is close to k_0 we choose k as independent variable and write

$$v = v_0 + \alpha(k-k_0) + \tfrac{1}{2}\beta(k-k_0)^2 + \ldots$$
where $\qquad\qquad\qquad\qquad\qquad\qquad\qquad\qquad\qquad\qquad (1.11)$

$$\alpha = \left[\frac{dv}{dk}\right]_0 \qquad \beta = \left[\frac{d^2v}{dk^2}\right]_0$$

the subscript zero indicating that the derivatives are evaluated at $k = k_0$. The main contribution to the integral (1.7) comes from k values around k_0, and $\delta = k - k_0$ is therefore small in this region. We put $k = k_0 + \delta$ and expand the term $2\pi i k(x - ut)$, remembering that $ku = v$, to obtain, with neglect of second-order terms,

$$2\pi i k\{x - u(k)t\} \simeq 2\pi i k_0(x - u_0 t) + 2\pi i \delta(x - \alpha t)$$

where u_0 is the phase velocity for wave number k_0. The integration (1.7) may then be performed and the result is

$$\psi(x, t) \simeq a\sqrt{\frac{\pi}{\sigma}}\exp\left(\frac{-\pi^2(x-\alpha t)^2}{\sigma}\right)\exp\{2\pi i k_0(x - u_0 t)\}. \qquad (1.12)$$

The complex factor indicates a wave train travelling with the mean phase velocity u_0. But the wave profile, arising from the amplitude factor, is evidently centred on the point $x = \alpha t$ and thus moves with a velocity α different from that of the constituent wave trains:

$$v_g = \alpha = \frac{dv}{dk} \qquad (1.13)$$

where the derivative is evaluated for the mean k value, $k = k_0$ (though we now drop the subscript), appropriate to the packet. The fact that the packet whose dispersion is determined by (1.10) does indeed move with the particle velocity, which appears in the de Broglie relationship (1.3), follows easily. For

$$v_g = \frac{dv}{dk} = \frac{1}{h}\frac{dE}{dk} = \frac{1}{h}\frac{d}{dk}\left(\frac{1}{2m}\,p^2 + v\right)$$

$$= \frac{1}{h}\frac{d}{dk}\left(\frac{h^2 k^2}{2m} + V\right) = \frac{hk}{m} = v.$$

Thus the packet actually stays with the particle whose behaviour it is supposed to describe.

There is, however, a difference between the behaviour of packets for photons and for material particles, in so far as the packet describing a material particle "spreads" as it travels. This result follows when the higher terms in (1.11) are taken into account, and will be considered again later. We also defer at this point any consideration of the meaning or "physical reality" of ψ in the two cases; it is here sufficient to accept ψ as a well-defined function which mathematically determines the behaviour of the particle.

It is apparent that there is a considerable similarity between the equations and concepts associated with photons and material particles. This close correspondence is indicated in Table 1.1, which summarizes the main features of the "duality" referred to at the beginning of this section.

Statistical interpretation of the wave function

Although the coexistence of wave and particle properties may now seem more acceptable, since the idea appears to resolve more difficulties than it creates, the question might still be asked: Are photons and electrons really waves or particles? The answer must be that they are neither: they are "quantum particles" which behave like classical particles under some conditions and like waves under others, though in all cases this behaviour may be mathematically described in terms of a wave function ψ. "Classical" behaviour is familiar to us from the study

TABLE 1.1

	Photon	Particle of mass m
Wave train Frequency and energy	$\exp\{2\pi i(kx - vt)\}$, $v = E/h$ v: frequency of radiation E: energy of photon	$\exp\{2\pi i(kx - vt)\}$, $v = E/h$ v: frequency in wave function (no direct physical meaning) $E = \frac{1}{2}mv^2 + V$: "classical" energy of particle

Phase velocity
de Broglie
  wave length
Group velocity
	$u = c$: velocity of light
	
	$\lambda = c/v = h/p$
	$v_g = c$: photon velocity (no
	  dispersion)
		$u = E/\sqrt{\{2m(E - V)\}} = E/mv$
		(unobservable)
		$\lambda = u/v = h/mv$
		$v_g = v$: particle velocity
		(dispersion, leading to
		"spreading" of packet)

of moving objects in the laboratory, but there is no compelling reason to expect that photons and electrons will behave in a similar way.

The extraordinary behaviour of quantum particles may be demonstrated by a simple diffraction experiment (Fig. 1.2) in which a beam of particles, each described by a plane wave or a packet of such waves (essentially homogeneous in wave length), is incident upon a screen containing slits L_1 and L_2. The form of the diffraction pattern produced on the plate P depends only on the *wave length* characterizing an incident particle, not on the intensity of the beam—measured by the amount of energy, or number of quantum particles, passing through unit cross-section per second. The remarkable fact is that the pattern persists even when the intensity is so low that there is never more than one particle between S and P at any given time. The value of $|\psi|^2$ at a point on the plate, calculated as in elementary physical optics, then certainly refers to only *one particle* and measures the *probability* that it will land at the given point: in a long exposure the same "experiment" is repeated by a very large number of particles and the whole pattern will appear. But there is no possibility of the interference being between *different* photons or electrons, passing through the different slits; the wave function ψ refers to *one particle* and simply provides the correct mathematical prescription for calculating where it is likely to be found in a given experiment.

The statistical interpretation of the wave function, due to Born, is a fundamental feature of quantum mechanics. It is possible in principle to calculate the wave function, from the equations of quantum mechanics, but from this function it is only possible to discuss the *probability* of finding a particle at any given point; the exact causality characteristic of classical physics no longer applies. Comprehensive discussions of hypothetical diffraction experiments are available elsewhere (e.g. Feynmann, 1965) and reveal that any successful attempt to obtain more definite information about the position of a particle in such an experiment (e.g. which slit it went through) would destroy the diffraction pattern. Such an experiment must, by definition, result in concentration of ψ within a small region of high probability, and the conditions of the original experiment therefore no longer obtain. It is now generally accepted that it is meaningful to talk about the "position" of a quantum particle only when its position has been ascertained between prescribed limits by means of an experiment and, more generally, that in any such experiment the actual state of the system may be grossly disturbed by interaction with the observer. In classical physics it is assumed that the

Fig. 1.2. Diffraction experiment. S is a source, L_1 and L_2 are two slits. A diffraction pattern (intensity indicated schematically) is formed at a photographic plate P.

interaction between system and observer may be made arbitrarily small: in quantum physics, where the processes considered involve particles as small as electrons and photons, this is not an assumption that can be consistently maintained. This point of view was first stressed by Heisenberg, and will be taken up in Chapter 4. Heisenberg and Schrödinger both succeeded in formulating equations to describe the behaviour of a quantum particle; Heisenberg's "matrix mechanics" and Schrödinger's "wave mechanics", later shown to be equivalent, are now regarded as part of the general discipline of "quantum mechanics".

The quantum theory of photons and the electromagnetic field turns out to be more involved and difficult than that of material particles, and lies outside the scope of this book. Fortunately, this development is not essential in many areas of quantum mechanics, particularly those concerned with the structure of matter and the behaviour of particles such as electrons and nuclei.‡ The wave function for such a particle then satisfies the wave equation first given by Schrödinger, which is "derived" by a simple argument in the next section. In the limit where relativistic effects are negligible, and where there is no radiation field, this equation appears to provide a completely adequate basis for discussing the motion of a particle; in this sense it is the quantum theory counterpart of Newton's equations—to which it must indeed be equivalent in the limit of large mass, where classical mechanics is known to be valid. It also turns out that interaction with an applied electromagnetic field, required for example in spectroscopic applications of the theory, may adequately be admitted in a semi-classical way and that the more important relativistic effects may be included by adding certain "spin" terms. The Schrödinger approach thus provides a satisfactory foundation for applications of quantum mechanics in many parts of physics and in almost the whole of chemistry.

1.2.   The Schrödinger equation for a particle

We now look for the equation whose solutions are wave functions of the kind discussed in Section 1.1, treating first the case of a particle of mass m moving in a straight line. The solutions must then be plane waves, or packets of plane waves, satisfying a one-dimensional equation (in, say, the x-coordinate). The equation we want cannot be the classical wave equation

$$\frac{\partial^2 \psi}{\partial x^2} = \frac{1}{u^2}\frac{\partial^2 \psi}{\partial t^2} \tag{1.14}$$

because the phase velocity already introduced is given by

$$u = \lambda v = \frac{E}{p}$$

and clearly involves dynamical variables (E and p); this would mean that the differential equation for ψ would depend on the energy and/or momentum in the particular state considered—and a mixture of *different* wave trains (as in a packet) could not be a solution of any *one* of these equations. The equation we want should involve only the universal constants such as e, m or h, and none of the dynamical variables such as energy or momentum.

Let us therefore work backwards, assuming (1.4) to represent a free electron travelling in the positive x direction with momentum p and

‡In low-energy processes a nucleus may be regarded as a single heavy particle.

kinetic energy E, and adopting the upper sign in (1.4) for reasons which will become clear presently. Thus,

$$\psi(x,\, t) = \exp\{2\pi ik(x-ut)\} = \exp\left(i\frac{p}{\hbar}x\right)\exp\left(-i\frac{E}{\hbar}t\right),\quad (1.15)$$

where we introduce the "rationalized" Planck constant $\hbar = h/2\pi$ in order to avoid the repeated appearance of factors of 2π. From (1.15) we obtain

$$\frac{\partial^2\psi}{\partial x^2} = -\frac{p^2}{\hbar^2}\psi,\quad \text{and}\quad \frac{\partial\psi}{\partial t} = -i\frac{E}{\hbar}\psi. \quad (1.16)$$

But for a free electron, the classical relationship

$$E = \frac{1}{2m}p^2$$

is known from experiment to be valid. This suggests that for a free electron the required differential equation is

$$i\hbar\frac{\partial\psi}{\partial t} = -\frac{\hbar^2}{2m}\frac{\partial^2\psi}{\partial x^2}. \quad (1.17)$$

This is in fact the one-dimensional form of Schrödinger's equation for a free electron.

The extension to the three-dimensional case is immediate. In this case the energy can be expressed as

$$E = \frac{1}{2m}(p_x{}^2 + p_y{}^2 + p_z{}^2)$$

and a similar argument shows that the derivatives satisfy the equation

$$i\hbar\frac{\partial\psi}{\partial t} = -\frac{\hbar^2}{2m}\left(\frac{\partial^2\psi}{\partial x^2} + \frac{\partial^2\psi}{\partial y^2} + \frac{\partial^2\psi}{\partial z^2}\right)$$

or, with the usual notation,

$$i\hbar\frac{\partial\psi}{\partial t} = \frac{\hbar^2}{2m}\nabla^2\psi. \quad (1.18)$$

This equation gives a satisfactory account of the behaviour of a *free particle* moving in three-dimensional space.

Schrödinger's equation

So far we have considered a *free* particle (no applied fields) whose energy is purely kinetic. We must now consider a particle moving in some external field (assumed here to depend only on position) corresponding to a classical potential energy $V(x, y, z)$. The complete energy expression is then

$$E = \frac{1}{2m} (p_x{}^2 + p_y{}^2 + p_z{}^2) + V(x, y, z).$$

If we assume the frequency factor is still determined by the energy, according to (1.16), and the wave length by the momentum, it seems plausible that the correct generalization of (1.18) will be

$$i\hbar \frac{\partial \psi}{\partial t} = -\frac{\hbar^2}{2m} \nabla^2 \psi + V\psi. \qquad (1.19)$$

This is the (time-dependent) *Schrödinger equation* for a particle and determines how any given state, represented by the wave function ψ, will develop in time. It serves, for example, to determine how a wave packet of the form (1.6) would be "scattered" on meeting a region of variable potential—perhaps due to a point charge or some kind of "potential barrier". Although the equation is here formulated in terms of a Cartesian coordinate system, its transformation to any other system (e.g. polar or cylindrical coordinates) is a purely mathematical exercise; $\nabla^2\psi$ is expressed in a general coordinate system in Appendix 1.

It is customary to write (1.19) in the form

$$\mathsf{H}\psi = i\hbar \frac{\partial \psi}{\partial t} \qquad (1.20)$$

where $\mathsf{H}\psi$ is an abbreviation for the right-hand side of (1.19), obtained by applying the *Hamiltonian operator*‡

$$\mathsf{H} = -\frac{\hbar^2}{2m} \nabla^2 + V \qquad (1.21)$$

to the wavefunction ψ. The properties of operators are discussed in Chapter 3; but here we use H merely as a convenient shorthand, $\mathsf{H}\psi$ (where ψ is any function) being interpreted simply by putting ψ on the right of each term in (1.21). It may be noted that (1.21) may be obtained from the classical energy expression§ in a formal way by replacing each momentum component by a differential operator, $p_x \rightarrow \mathsf{p}_x$, etc., where

$$\mathsf{p}_x = \frac{\hbar}{i} \frac{\partial}{\partial x}, \quad \mathsf{p}_y = \frac{\hbar}{i} \frac{\partial}{\partial y}, \quad \mathsf{p}_z = \frac{\hbar}{i} \frac{\partial}{\partial z} \qquad (1.22)$$

‡Throughout this book, a distinctive type (Gill Sans) will be used to indicate operators.
§The expression for the energy in terms of *momentum* components (not velocities) and coordinates is called the Hamiltonian function (see, for example, Pauling and Wilson, 1935); we write $E = H(p_x, p_y, p_z, x, y, z)$. The equation preceding (1.19) is of this form, and the prescription (1.22) at once yields (1.19).

and noting that $p_x{}^2 = p_x p_x = -\hbar^2(\partial^2/\partial x^2)$, etc. This particular association of differential operators with dynamical variables is characteristic of Schrödinger's formulation of quantum mechanics.

The time-independent equation

In many cases we are concerned with an important special class of solutions of (1.19), in which ψ has the form

$$\psi(x, y, z, t) = \phi(x, y, z)f(t)$$

or, to use a "vector" notation in which **r** stands for the set of coordinates x, y, z,

$$\psi(\mathbf{r}, t) = \phi(\mathbf{r})f(t). \tag{1.23}$$

On substituting (1.23) into (1.19) and dividing by ψ, we obtain

$$i\hbar \frac{1}{f(t)} \frac{df(t)}{dt} = \frac{1}{\phi(\mathbf{r})} \left(-\frac{\hbar^2}{2m} \nabla^2 \phi(\mathbf{r}) + V(\mathbf{r})\phi(\mathbf{r}) \right).$$

But the two sides of this equation depend on quite different sets of variables, $\mathbf{r} = (x, y, z)$ and t, respectively. In order to satisfy the equation for all values of t, when x, y, z are given (fixing the right-hand side), it is clear that the left-hand side must be a constant; and the right-hand side must likewise be a constant—with the same value. If the constant is denoted by E this means

$$\frac{df}{dt} = -i\frac{E}{\hbar}f$$

and hence

$$f(t) = \exp\left(-i\frac{E}{\hbar} t \right).$$

To identify E, which clearly has the dimensions of energy, we note that for a free particle the solution must reduce to (1.15); in this case, and it will be assumed in general, E is the *energy of the system* in the state with wave function (1.23). We also note that the ambiguity in sign which occurs in the travelling wave solution (1.4) of the classical wave equation has been removed by going over to (1.19) which contains only a *first* derivative with respect to time.

The remaining equation which must be satisfied if (1.19) is to admit a solution of the special form (1.23) follows on equating to E the *right*-hand side of the equation following (1.23). Thus, with the notation (1.21),

$$H\phi = E\phi. \tag{1.24}$$

This is usually referred to as the "time-independent Schrödinger equation" to distinguish it from (1.20). It is clear that from any solution $\phi(\mathbf{r})$ of the time-independent Schrödinger equation, we can construct a particular solution of the time-dependent Schrödinger equation $\psi(\mathbf{r}, t)$, of the form

$$\psi(\mathbf{r}, t) = \phi(\mathbf{r}) \exp(-iEt/\hbar). \tag{1.25}$$

These particular solutions, and the time-independent Schrödinger equation (1.24) which determine them, play a dominant role in the quantum theory of valency and molecular structure. They represent *stationary states*, whose physical meaning will be discussed in the next section. Solutions of the time-dependent Schrödinger equation (1.20) are required mainly in the treatment of scattering and similar processes; but even in this case it is usual to start from the time-independent equation (1.24), obtaining special solutions of the form (1.25) which can then be used to build up a more general solution—just as plane waves may be used to build up a general wave packet. Equation (1.24) is therefore absolutely fundamental to all that follows.

The preceding formulation of Schrödinger's two equations is based merely on a plausibility argument. That the equations are correct can be determined only by exhaustive applications; but the evidence that (1.20) and (1.21) are satisfactory, in all situations where relativistic effects are negligible, is by now overwhelming. We note that effects due to an external magnetic field (which introduces velocity dependent terms into the potential energy) have not yet been considered, but that their inclusion (Vol. 2)‡ presents no difficulty in non-relativistic theory.

1.3. Probability density and probability current

At this point it is necessary to consider more fully the statistical meaning of $|\psi|^2$ and to ask what mathematical restrictions must be imposed on the function ψ. In physics, the general solution of a partial differential equation is seldom required; arbitrary functions must usually be eliminated in order to obtain a particular solution which is physically acceptable and conforms to any given boundary conditions. In wave mechanics such restrictions arise mainly from the physical nature of probability functions.

Probability density functions

The Schrödinger equations (1.20) and (1.24) are "linear", in the sense that if ψ is any solution then so must be $c\psi$, where c is an arbitrary constant.§ The solution is thus arbitrary to within a constant factor, and in quantum theory we must take ψ and $c\psi$ ($c \neq 0$) to

‡References are to other volumes in Topic 2.
§ c is commonly used to denote a numerical constant when there is no risk of configuration with the velocity of light.

represent one and the same state. The statistical interpretation of $|\psi|^2$, however, allows us to introduce a convention which effectively removes this ambiguity. Given any solution, we multiply by a constant so chosen that $|\psi(\mathbf{r}, t)|^2 d\mathbf{r}$ is the *absolute* (rather than a relative) probability of finding a particle in volume element $d\mathbf{r}$ at point $\mathbf{r}$ and at time t. Here we continue to use the "vector" notation in which $\mathbf{r}$ and $d\mathbf{r}$, respectively, stand for the variables defining position (e.g. x, y, z) and for the volume element (e.g. $dx\,dy\,dz$). The function $|\psi|^2$ is then a *probability density function*. Since the probability of finding the particle *somewhere* in space is unity, we must require

$$\int |\psi|^2 d\mathbf{r} = 1 \tag{1.26}$$

where the integration is carried out over the whole of space. The function ψ is in this case said to be *normalized*. An alternative solution $c\psi$ will then no longer be acceptable unless c happens to be a uni-modular complex number of the form $e^{i\theta}$ where θ is an arbitrary constant: such a "phase factor" has no physical meaning and may generally be omitted.

The significance of the "stationary states" with wave functions of the form (1.25) is now clear. Such a state is "stationary" in the sense that

$$|\psi(\mathbf{r}, t)|^2 = |\phi(\mathbf{r})|^2 \tag{1.27}$$

where the right-hand side is time-independent. This means that the probability density function, which describes where the particle is likely to be found, persists indefinitely without change. The probability $|\phi(\mathbf{r})|^2$ is in principle physically observable, being the fractional number of times the particle would be found in $d\mathbf{r}$, at point $\mathbf{r}$, if observations were made a very large number of times under identical conditions.[‡] The normalization condition for a stationary state reduces to

$$\int |\phi(\mathbf{r})|^2 d\mathbf{r} = 1$$

as follows at once from (1.26) and (1.27).

The fact that the wave function should be normalizable is a basic physical requirement, which is usually put in a slightly different way by saying that any physically acceptable solution of the wave equation, even before normalization, must satisfy the condition[§]

$$\int |\psi|^2 d\mathbf{r} = \text{finite.} \tag{1.28}$$

Functions with this property are said to be of *integrable square*, or to be

---

‡In practice, of course, the form of a probability density function must be inferred less directly (e.g. from X-ray scattering experiments on a large assembly of particles).

§We sometimes use the same symbol ψ for a function which we have not troubled to normalize (or whose normalization presents some difficulty); the sense will be clear from the context.

quadratically integrable. It is possible that ψ may become infinite at a finite number of points (singularities) but it must do so in such a way that the integral converges. We also note that since the probability of finding the particle somewhere in space must be unity *at all times* it is necessary that the normalization integral (1.28) be time-independent.

Another obvious physical requirement is that the square of the wave function should be *single valued*, for it would be nonsense to admit two distinct values of the probability of finding the particle in a given place at a given time. This requirement is usually taken to imply that ψ itself is single-valued, though justification of the stronger condition is not immediate.

Probability current. Definition and Implications

To investigate other restrictions on the wave function it is necessary to consider how the value of $|\psi|^2$ at a given point may change with time. This leads to the concept of *probability current*. The probability distribution may be compared with unit mass of a compressible fluid; it may be spread over a large region with low density or it may be compressed into a small region at high density, conservation of total "mass" requiring that

$$\int \rho \, d\mathbf{r} = 1$$

where in the present instance $\rho = |\psi|^2$. If the "mass" within a given region (i.e. the integral of ρ over the region) is changing, then "mass" must be flowing across the boundary and we can define a "mass current density" as the rate of flow across unit area normal to the direction of flow. In our case the "mass" is a probability (integral of ρ over some region), the (mass) density is the probability density $\rho = |\psi|^2$, and we shall define a *probability current density* exactly analogous to the mass current density. The meaning is clear: if there is a net probability current out of any closed region, the chances of finding the particle within that region must be diminishing.

To simplify the argument we consider the one-dimensional situation‡ in Fig. 1.3, the elementary slab shown having as its surfaces planes on which ψ is constant, so that $\psi = \psi(x)$. If the slab has unit surface area the probability of finding the particle within the element is $\psi^*\psi \, dx$ and this must be changing at a rate $J_x(x) - J_x(x + dx)$. Equating the time rate of change of $\psi^*\psi \, dx$ to the difference of currents, we have

---

‡This might appear inconsistent with the quadratic integrability of $|\psi|^2$ since ψ is constant in the y and z direction: the argument is, however, applied only over a region so small that the y, z dependence may be ignored.

$$\frac{\partial \rho}{\partial t} = \frac{\partial}{\partial t}(\psi^*\psi) = -\frac{\partial J_x}{\partial x} \tag{1.29}$$

which really amounts merely to a definition of the current in the x direction. To identify J_x in terms of the wave function we write the time derivative as

$$\frac{\partial}{\partial t}(\psi^*\psi) = \frac{\partial \psi^*}{\partial t}\psi + \psi^*\frac{\partial \psi}{\partial t}$$

and then use (1.19) and its complex conjugate which gives $\partial \psi^*/\partial t$. We

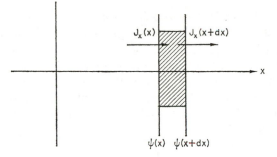

FIG. 1.3 Definition of the probability current density. The rate of increase of probability of finding the particle in the shaded region (an elementary slab of unit-cross section) is $J_x(x) - J_x(x+dx)$.

assume here that V is real and velocity independent (magnetic effects excluded) and the expression on the right reduces to

$$\frac{\hbar}{2im}\left[\frac{\partial^2 \psi^*}{\partial x^2}\psi - \psi^*\frac{\partial^2 \psi}{\partial x^2}\right]$$

which may be written as an x derivative to give

$$\frac{\partial}{\partial t}(\psi^*\psi) = \frac{\hbar}{2im}\frac{\partial}{\partial x}\left[\frac{\partial \psi^*}{\partial x}\psi - \psi^*\frac{\partial \psi}{\partial x}\right].$$

By comparison with (1.29) we obtain

$$J_x = \frac{\hbar}{2im}\left[\psi^*\frac{\partial \psi}{\partial x} - \psi\frac{\partial \psi^*}{\partial x}\right] \tag{1.30}$$

as the probability current density in the x direction.

In three dimensions, the derivation is entirely similar; the probability current has three components, each given by an expression of the form (1.30). If we use the subscripts 1, 2, 3 to denote the x, y, z

components of the position vector **r**, replacing (x, y, z) by (r_1, r_2, r_3), the results become

$$J_\alpha = \frac{\hbar}{2im}\left(\psi^* \frac{\partial \psi}{\partial r_\alpha} - \psi \frac{\partial \psi^*}{\partial r_\alpha}\right) \quad (\alpha = 1, 2, 3) \tag{1.31}$$

and the equation giving the rate of change of probability density at any point is, by analogy with (1.29)

$$\frac{\partial \rho}{\partial t} = -\sum_\alpha \frac{\partial J_\alpha}{\partial r_\alpha}. \tag{1.32}$$

This equation is frequently written in vector notation. If **J** is the vector with components J_α, the right-hand sum is recognized as div **J** and (1.32) can be written in the familiar form of an "equation of continuity":

$$\frac{\partial \rho}{\partial t} + \text{div } \mathbf{J} = 0. \tag{1.33}$$

Similarly, equations (1.31) may be collected into the form

$$\mathbf{J} = \frac{\hbar}{2im} (\psi^* \text{ grad } \psi - \psi \text{ grad } \psi^*). \tag{1.34}$$

In practice, of course, the probability current is actually evaluated by using the component form (1.31). Again we recall that the equations have been derived assuming no external magnetic field. The modifications necessary when a magnetic field is applied are considered elsewhere (Vol. 2).

After this digression the nature of the remaining conditions to be imposed on ψ becomes clear. These conditions relate to continuity. If ψ were discontinuous across some surface (i.e. if a finite change occurred in an infinitesimally small region), the derivatives and hence the current would become *infinite*. This possibility must be ruled out because if a region were chosen with boundaries arbitrarily close to such a surface the probability of finding the particle within that region would increase at an infinite rate.

If the *gradient* were discontinuous across some surface, the second derivative would become infinite. The Schrödinger equation (1.20) or (1.24) could not then be satisfied unless the potential were infinite on the surface. We thus require that the gradient of a wave function be continuous everywhere except where the potential becomes infinite.

The mathematical restrictions suggested by such considerations are collected below and are adopted in all that follows:

(i) ψ must be of integrable square;‡

(ii) the integral of $|\psi|^2$ over all space must be time-independent;

(iii) ψ must be single-valued;

(iv) ψ must be continuous;

(v)   the gradient of ψ must be continuous except where the potential becomes infinite.

Functions satisfying such requirements comprise a "class". For brevity, functions which are quadratically integrable, etc., are often referred to as functions "of class Q" or "of class L^2" a terminology which we shall often find useful.

We are now in a position to consider and solve a number of simple problems which will give considerable insight into the behaviour of quantum mechanical systems.

1.4.   The classical limit for motion of a wave packet

Before leaving this brief discussion of the origins of quantum mechanics we should convince ourselves that classical dynamics is included as a suitable limiting case, since there is no doubt that particles visible to us in the laboratory move according to the Newtonian law:

Rate of change of linear momentum = Applied force.

The distinctive feature of the classical situation is that the particle position and the applied force may be measured with high accuracy; in other words, classical dynamics should apply in the limit where a wave packet is so compact that there is negligible probability of finding the particle at an appreciable distance from the centroid of the packet.

We therefore ask whether a Newtonian law applies to the *average* values of force and momentum associated with a particle described by a wave packet. Let us consider motion along the x-axis. The average x-coordinate of the particle, described by probability density $\rho = \psi^*\psi$, is

$$\bar{x} = \int x\psi^*\psi \, d\mathbf{r}, \tag{1.35}$$

i.e. the x-component of the centroid of the density. The x-component of the force acting, corresponding to the classical potential energy function $V = V(\mathbf{r})$, is $F_x = -(\partial V/\partial x)$ and when this is averaged over the wave packet we have

---

‡There is an apparent exception in the case of a particle free to travel throughout infinite space (e.g. represented by a plane wave); the interpretation of quadratic integrability in this case is dealt with later.

$$\overline{F}_x = -\int(\partial V/\partial x)\psi^*\psi\,d\mathbf{r}. \tag{1.36}$$

Newtonian dynamics will thus apply in the limiting case of strongly localized wave packets if we can show that, when the wave function develops in time according to (1.19), the particle moves in such a way that

$$\frac{d(m\bar{x})}{dt} = \bar{F}_x. \tag{1.37}$$

The proof that this equation holds was first given by Ehrenfest.

Here we sketch the argument, which should be completed by the reader. Since x is merely an integration variable in (1.35) we obtain

$$\frac{d}{dt}(m\bar{x}) = \int x\,\frac{\partial}{\partial t}(\psi^*\psi)\,d\mathbf{r} = -\frac{i\hbar}{2}\int x\,\frac{\partial}{\partial x}\!\left(\frac{\partial\psi^*}{\partial x}\,\psi - \psi^*\,\frac{\partial\psi}{\partial x}\right)dx\,dy\,dz,$$

where the $\partial(\psi^*\psi)/\partial t$ has been taken from the derivation of the current density (p. 16). On integrating by parts and assuming that ψ and its derivatives vanish at infinity we obtain

$$\frac{d}{dt}(m\bar{x}) = i\hbar\int \psi^*\frac{\partial\psi}{\partial x}\,dx\,dy\,dz.$$

The rate of change of this quantity is

$$\frac{d^2}{dt^2}(m\bar{x}) = i\hbar\int\left[\frac{\partial\psi^*}{\partial t}\frac{\partial\psi}{\partial x} + \psi^*\frac{\partial}{\partial t}\!\left(\frac{\partial\psi}{\partial x}\right)\right]dx\,dy\,dz.$$

For functions satisfying the required conditions (p. 18) the order of differentiations in the second term may be reversed. We then substitute for $\partial\psi/\partial t$ and $\partial\psi^*/\partial t$ from (1.19) and its complex conjugate, which completely determine the time-development of the packet: there is a cancellation of the terms involving ∇^2 and further integration by parts yields the final result

$$\frac{d^2(m\bar{x})}{dt^2} = -\int \psi^*\psi\left(\frac{\partial V}{\partial x}\right)dx\,dy\,dz.$$

Comparison with (1.36) then establishes the equation of motion (1.37). Similar results hold, of course, for the y- and z-components.

These results are truly remarkable. They show that the three-dimensional motion of a quantum particle, described by a wave packet such as that shown in Fig. 1.1 and travelling in an arbitrary potential field, will coincide with that predicted by Newtonian dynamics provided the packet is sufficiently localized—in spite of the complete dissimilarity of the methods of calculation. We know that such packets tend to "spread" (p. 6) and that the precision with which the classical picture applies will therefore diminish as we continue to make observations; but we shall find later that for sufficiently massive particles the rate of spreading is so exceedingly small that deviations from Newtonian behaviour would not be detected during the time of an experiment. For many purposes, nuclei may be treated as classical particles; but for electrons this is seldom possible and we have no alternative but to start from Schrödinger's equations. The implications of wave-packet localization are discussed in later sections. Here we note only that the vertical coincidence of the laws of motion for classical and quantum particles, in the limit of large mass, is an empression of the famous *correspondence principle* advanced by Bohr two years before Schrödinger's successful formulation of the quantum equations of motion embodied in (1.19).

REFERENCES

FEYNMAN, R. P., LEIGHTON, R. B. and SANDS, M. (1965) *The Feynman Lectures in Physics,* Vol. III, Addison-Wesley, Reading, Mass.
PAULING, L. and WILSON, E. B. (1935) *Introduction to Quantum Mechanics,* McGraw Hill, New York.

CHAPTER 2

SOME SIMPLE SOLUTIONS OF SCHRÖDINGER'S EQUATION

2.1. The particle in a container

The aim of this chapter is to present and discuss some simple solutions of the time-independent Schrödinger equation (1.24), using a minimum of mathematics. In this first example we consider the nature of the solutions for a particle confined to a certain region of space, in which the classical potential energy function has the one-dimensional form shown in Fig. 2.1, and then work out the solution for a special kind of three-dimensional "container" or "box" in which the potential V rises to infinity at the boundaries.

Fig. 2.1. Energy diagram for one-dimensional motion of a particle, (a) showing classical limits of motion corresponding to energy E, (b) showing simplified "potential box" with same limits (note that the zero of potential energy is arbitrary and has been taken as $V = 0$ inside the box).

First we recall the classical description of the (one-dimensional) motion of a particle whose potential energy varies as in Fig. 2.1a. The total energy E is the sum of kinetic and potential terms, T and V,

$$E = T + V,$$

where T, from its nature, is essentially a *positive* quantity. If a horizontal line is drawn at a height representing the total energy E (Fig. 2.1a), the breakdown into these two parts is pictorially obvious and it is clear that the particle is bound within the region between the vertical lines—within which T is positive. The balance between potential and kinetic energy oscillates during the motion, with $T \to 0$ as the particle approaches the boundaries of the "container". If the particle is given more energy, the effective size of the container increases, and if E is so high as not to intersect V then the particle is no longer bound but becomes free and may travel infinitely far in either direction. To approximate the situation described in Fig. 2.1a it is frequently useful to use the "potential box" (Fig. 2.1b) to define the region in which the particle is confined.

We now look for the main features of a stationary state of a particle in the box of Fig. 2.1b, according to quantum mechanics. The Schrödinger equation for such a state is (1.24) and takes different forms inside and outside the box:

$$\textit{Inside the box}: \qquad -\frac{\hbar^2}{2m}\frac{\partial^2 \phi}{\partial x^2} = E\phi.$$

$$\textit{Outside the box}: \qquad -\frac{\hbar^2}{2m}\frac{\partial^2 \phi}{\partial x^2} = -(V_0 - E)\phi.$$

The solution of the first equation is sinusoidal‡: in general, with arbitrary constants A and B,

$\textit{Inside}: \qquad \phi = A \sin kx + B \cos kx \qquad k = \sqrt{(2mE/\hbar^2)}.$

$\textit{Outside}: \qquad \phi = C e^{k'x} + D e^{-k'x} \qquad k' = \sqrt{\{2m(V_0 - E)/\hbar^2\}}.$

To get a physically satisfactory solution the constants must be so chosen in the three regions (one inside and two outside), that its constituent parts join smoothly and satisfy the other necessary conditions (p. 18). The exponential terms which *increase* on going out from the box must therefore be discarded (i.e. be given zero coefficients) and a properly matched solution might take the form Fig. 2.2. In fact, this matching can be achieved only for certain special values of E, which determine both the wave length inside the box and the rapidity with which the exponential "tails" fall off. This means that stationary states of the system can exist *only for certain values of the energy*. The existence of "quantized" energy levels is, of course, an important distinctive feature of the quantum mechanical description. A second important non-classical feature is the possibility of "penetration" of

---

‡In some cases the complex exponential form $\phi = A' e^{ikx} + B' e^{-ikx}$ is more convenient.

the particle into regions where it would not be expected, namely beyond the boundaries of the box where the exponential tails of the wave function may persist for some distance.

The solution of this simple box problem, surprisingly perhaps, cannot be completed without recourse to numerical or graphical methods to evaluate the constants A, B, etc., but the form of the wave function suggests how the boundary conditions may be simplified so as to make solution easier. Thus as V_0 is taken larger and larger the exponential decrease of ϕ outside the box will become more and more rapid (k' large): in the limit when $V_0 \to \infty$ we speak of an "infinitely deep" potential box and may take $\phi = 0$ *at the boundaries of the box*.

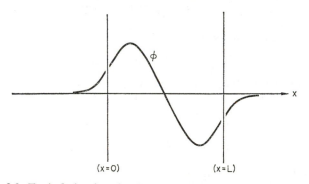

Fig. 2.2. Typical eigenfunction for a one-dimensional box. The vertical lines indicate the box boundaries, at which the external and internal solutions must join smoothly.

Let us turn at once to the three-dimensional case of a box in the form of a cube, outside which the potential energy function V rises to an indefinitely large value. Such a box provides a useful "model" for various types of real system (e.g. a gas molecule in a container, or an electron in a metal). If the length of each edge is L, and the zero of potential energy (which is always arbitrary) is chosen so that $V = 0$ inside the cube, the Schrödinger equation (1.24) becomes

$$\mathsf{H}\phi = -\frac{\hbar^2}{2m}\left(\frac{\partial^2\phi}{\partial x^2}+\frac{\partial^2\phi}{\partial y^2}+\frac{\partial^2\phi}{\partial z^2}\right) = E\phi. \tag{2.1}$$

It is useful to look first for a particular type of solution (cf. p. 12) of the form

$$\phi(x, y, z) = X(x)\,Y(y)\,Z(z). \tag{2.2}$$

This technique is known as "separation of the variables" and is discussed in Appendix 2.

On substituting (2.2) into (2.1), and dividing both sides by ϕ, we obtain

$$-\frac{\hbar^2}{2m}\left(\frac{1}{X}\frac{\partial^2 X}{\partial x^2}+\frac{1}{Y}\frac{\partial^2 Y}{\partial y^2}+\frac{1}{Z}\frac{\partial^2 Z}{\partial z^2}\right)=E. \qquad (2.3)$$

The first, second and third terms in the left-hand side depend solely on x, y and z respectively. In order that the sum of these three terms may be equal to a constant E for any values of x, y and z, each of these terms must be a constant—for otherwise each would vary *independently* of the others. We call these constants a, b and c, respectively, and have

$$\frac{1}{X}\frac{d^2 X}{dx^2}=a, \quad \frac{1}{Y}\frac{d^2 Y}{dy^2}=b, \quad \frac{1}{Z}\frac{d^2 Z}{dz^2}=c, \qquad (2.4)$$

where the constants must satisfy

$$E=-(\hbar^2/2m)(a+b+c). \qquad (2.5)$$

The solutions of the above equations were considered at the beginning of this section: thus, for the first equation in (2.4),

$$X(x)=A\sin(x\sqrt{-a})+B\cos(x\sqrt{-a}) \quad (a<0),$$
$$X(x)=C\exp(x\sqrt{a})+D\exp(-x\sqrt{a}) \quad (a>0), \qquad (2.6)$$

where A, B, C and D are arbitrary constants.

To fix the constants in (2.6), we use the boundary condition $\phi=0$ corresponding to an effectively infinite rise in potential energy—which implies simply that the particle cannot penetrate the walls. A corner of the cube is taken as the origin of the coordinate system and the boundary conditions on $X(x)$ become $X=0$ at $x=0$ and $x=L$. When $a>0$ these conditions require, respectively,

$$C+D=0,$$
$$C\exp(L\sqrt{a})+D\exp(-L\sqrt{a})=0$$

and are satisfied only when $C=D=0$. This means that $X(x)$ is identically zero and the solution is of no physical interest.

When $a<0$, the boundary condition at $x=0$ tells us that $D=0$, while the condition at $x=L$ becomes

$$L\sqrt{-a}=n_x\pi, \quad n_x=1,2,3.\ldots \qquad (2.7)$$

Consequently, there exist non-vanishing solutions $X(x)$ only for values of a given by (2.7), and these are

$$X(x)=C\sin(n_x\pi x/L). \qquad (2.8)$$

The solutions for the y- and z-components are of exactly the same type and it follows from (2.2) that

$$\phi_{n_x n_y n_z} = N \sin (n_x \pi x/L) \sin (n_y \pi y/L) \sin (n_z \pi z/L), \quad \text{(a)}$$

$$E_{n_x n_y n_z} = \frac{\hbar^2 \pi^2}{2mL} (n_x^2 + n_y^2 + n_z^2), \quad \text{(b)} \quad (2.9)$$

where N is a normalization constant and is easily found to be $(2/L)^{3/2}$.

The energy values for which non-vanishing solutions exist are called energy *eigenvalues* and the corresponding wave functions are energy *eigenfunctions*. The eigenvalues in this example do not assume continuously variable values and are said to be *discrete*; they each depend upon three *quantum numbers*, n_x, n_y, n_z, which are used to label the corresponding states.

The state with the lowest possible energy is usually referred to as the ground state, $n_x = n_y = n_z = 1$. The energy and the wave function are then

$$E_{111} = \frac{3\hbar^2 \pi^2}{2mL^2}, \quad \phi_{111}(x, y, z) = \left(\frac{2}{L}\right)^{3/2} \sin \frac{\pi x}{L} \sin \frac{\pi y}{L} \sin \frac{\pi z}{L}$$

respectively. We note that ϕ_{111} does not become zero except at the walls; in other words there is no node inside the box. There is in fact a general theorem to the effect that the lowest energy eigenfunction of an equation such as (2.1) has no nodes.

The next lowest energy is obtained when one of n_x n_y and n_z is 2, the others each being 1. The energy is then

$$E_{211} = \frac{6\hbar^2 \pi^2}{2mL^2}$$

and there are *three* eigenfunctions with this same energy value:

$$\phi_{211}(x, y, z) = \left(\frac{2}{L}\right)^{3/2} \sin (2\pi x/L) \sin (\pi y/L) \sin (\pi z/L),$$

$$\phi_{121}(x, y, z) = \left(\frac{2}{L}\right)^{3/2} \sin (\pi x/L) \sin (2\pi y/L) \sin (\pi z/L),$$

$$\phi_{112}(x, y, z) = \left(\frac{2}{L}\right)^{3/2} \sin (\pi x/L) \sin (\pi y/L) \sin (2\pi z/L),$$

When, as in this case, two or more different eigenfunctions have the same eigenvalue the states and the energy level are said to be *degenerate* and the number of corresponding eigenfunctions is the *degree of degeneracy*. In this example, the first level above the ground state—the first "excited" state—has a three-fold degeneracy. We also note that each of ϕ_{211}, ϕ_{121} and ϕ_{111} has one nodal surface in the box, at $x = L/2$ $y = L/2$ and $z = L/2$ respectively. This is a general feature of wave functions; the higher-energy eigenfunctions contain a higher number of nodes.

A detailed discussion of the forms of the wave functions for higher quantum numbers, and of the correspondence between classical and quantum descriptions, is available elsewhere (e.g. Pauling and Wilson, 1935). Here we remark only that the characteristic quantization of the energy levels becomes less and less important in situations where classical mechanics would be expected to apply. Thus for a gas molecule in a 1-centimetre cube, with energy $\sim (3/2)kT$, the quantum numbers are generally extremely large numbers; discrete "jumps" on adding energy are then not noticeable, adjacent levels being separated by a very small fraction (usually $< 10^{-10}$) of the energy itself. The system appears to behave classically, as if continuous variation of the energy were permitted. For an electron, on the other hand, confined in a box comparable with the experimentally inferred size of the hydrogen atom, the energy difference between two adjacent levels may be of the same order as the energy itself, and it is obvious that classical mechanics is entirely inapplicable. Again (Section 1.4) classical mechanics appears as an appropriate limiting case.

Finally, it must be emphasized that the boundary conditions played an essential role in determining the eigenvalues and eigenfunctions. An alternative choice, which will in fact be given a physical interpretation in a later section, imposes a periodicity on the wave function by requiring, for example,

$$X(x = 0) = X(x = L)$$

$$\left(\frac{dX}{dx}\right)_{x=0} = \left(\frac{dX}{dx}\right)_{x=L} \tag{2.10}$$

and instead of (2.7) and (2.8), we then obtain

$$L\sqrt{-a} = 2n_x\pi, \quad n_x = 0, \pm 1, \pm 2, \ldots$$

$$X(x) = C \sin (2\pi n_x/L)x + D \cos (2\pi n_x/L)x. \tag{2.11}$$

The energy expression is obtained as before. The ground state in this case corresponds to $n_x = n_y = n_z = 0$, so that $E_{000} = 0$ and $\phi_{000} = \text{constant} = L^{-3/2}$. All the excited states are degenerate, the energy depending only on the squares of n_x, n_y, n_z which now take both positive and negative values.

2.2. The harmonic oscillator

For simplicity, we consider the one-dimensional harmonic oscillator. This corresponds to the classical situation in which a particle is attracted towards the origin by a force proportional to its displacement from the origin. If we denote the proportionality constant by $k(k > 0)$, the force and potential energy are, respectively,

$$F(x) = -kx, \quad V(x) = \tfrac{1}{2}kx^2$$

and the Schrödinger equation for the particle becomes

$$\mathsf{H}\phi = -\frac{\hbar^2}{2m}\frac{d^2\phi}{dx^2} + \tfrac{1}{2}kx^2\phi = E\phi. \tag{2.12}$$

To reduce this equation to a neater form, we introduce

$$\lambda = \frac{2m}{\hbar^2}E, \quad \alpha^2 = \frac{mk}{\hbar^2}$$

and obtain instead of (2.12)

$$\frac{d^2\phi}{dx^2} + (\lambda - \alpha^2 x^2)\phi = 0. \tag{2.13}$$

This equation should hold for all values of x, and it is useful to consider first how the solution must behave when $|x|$ is very large. The term $\lambda\phi$ in (2.13) can then be neglected in comparison with $-\alpha^2 x^2\phi$ and the equation reduces to

$$\frac{d^2\phi}{dx^2} = \alpha^2 x^2\phi.$$

Now it is clear that a trial function

$$X(x) = \exp\left(\pm\tfrac{1}{2}\alpha x^2\right)$$

would *almost* satisfy this equation; for, differentiating twice, we obtain

$$\frac{d^2X}{dx^2} = \pm\alpha \exp\left(\pm\tfrac{1}{2}\alpha x^2\right) + \alpha^2 x^2 \exp\left(\pm\tfrac{1}{2}\alpha x^2\right)$$

and if we are considering the case where $|x|$ is very large, then the first term on the right-hand side may be neglected by comparison with the second. We therefore conclude that the solution $\phi(x)$ of (2.13) resembles $X(x)$ when $|x|$ is large, and notice also that the negative sign in the exponent is required if $\int|\phi|^2 dx$ is to be finite.

When $|x|$ is not so large, $\phi(x)$ will be different from $X(x)$, but the difference may be expressed in terms of an unknown function which we shall call $u(x)$:

$$\phi(x) = u(x) \exp\left(-\tfrac{1}{2}\alpha x^2\right).$$

On putting this expression into (2.13), it appears that $u(x)$ must satisfy

$$\frac{d^2u}{dx^2} - 2\alpha x\, \frac{du}{dx} + (\lambda - \alpha)u = 0.$$

This is closely related to a well-known second-order differential equation: to obtain the standard form we change the independent variable from x to $\xi = x\sqrt{\alpha}$ and obtain, putting $u(x) = y(\xi)$,

$$\frac{d^2y}{d\xi^2} - 2\xi\, \frac{dy}{d\xi} + \left(\frac{\lambda}{\alpha} - 1\right)y = 0. \tag{2.14}$$

This equation may be solved by the standard method, discussed in Appendix 3, of writing $u(\xi)$ as a power series in ξ and determining the constant coefficients.

The results may be summarized as follows. In general the power series defining $u(\xi)$ is infinite and its value for ξ large rises even more rapidly than $\exp\left(\alpha x^2/2\right)$; the solution cannot then be normalized and must be rejected. On the other hand, for the *particular* λ values given by

$$\lambda = (2n+1)\alpha, \quad n = 0, 1, 2, \ldots \tag{2.15}$$

the series for $u(\xi)$ will terminate, becoming a finite polynomial of degree n. The corresponding function ϕ will be normalizable and therefore an allowed wave function. The finite polynomial of nth degree which satisfies (2.14) with λ given by (2.15) is the *Hermite polynomial, $H_n(\xi)$* (Appendix 3). The permissible stationary state energies of the harmonic oscillator may thus be written (remembering the definitions of λ, α)

$$E_n = (n + \tfrac{1}{2})\hbar\sqrt{(k/m)}. \tag{2.16}$$

Since the vibration frequency according to classical mechanics is given by

$$\nu = \frac{1}{2\pi}\sqrt{\frac{k}{m}} \tag{2.17}$$

it is customary to write (2.16) in the form

$$E_n = (n + \tfrac{1}{2})h\nu. \tag{2.18}$$

The corresponding wave functions, after normalization, are found to be

$$\phi_n(x) = \frac{1}{\sqrt{2^n n!}}\left(\frac{\alpha}{\pi}\right)^{\frac{1}{4}} H_n(x\sqrt{\alpha}) \exp\left(-\tfrac{1}{2}\alpha x^2\right). \tag{2.19}$$

The energies and wave functions for the ground state and first two excited states are given below:

$$E_0 = \tfrac{1}{2}hv \quad \phi_0 = (\alpha/\pi)^{1/4} \exp(-\tfrac{1}{2}\alpha x^2),$$

$$E_1 = \tfrac{3}{2}hv \quad \phi_1 = \sqrt{2}(\alpha/\pi)^{1/4}x\sqrt{\alpha} \exp(-\tfrac{1}{2}\alpha x^2),$$

$$E_2 = \tfrac{5}{2}hv \quad \phi_2 = \tfrac{1}{2}\sqrt{2}(\alpha/\pi)^{1/4}(2\alpha x^2 - 1) \exp(-\tfrac{1}{2}\alpha x^2),$$

respectively. The shapes of these eigenfunctions are shown in Fig. 2.3 which illustrates the general fact that ϕ_n contains n nodes, the energy

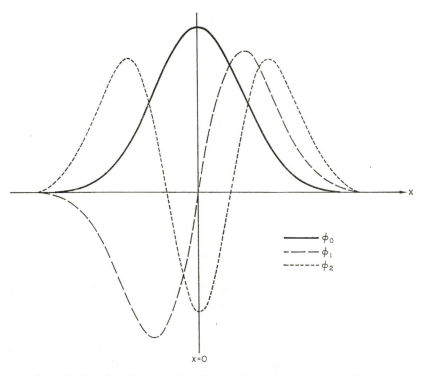

FIG. 2.3. The first three eigenfunctions of the linear harmonic oscillator. ϕ_0, ϕ_1, ϕ_2 correspond to quantum number $n = 0, 1, 2$.

again increasing with n. In Fig. 2.4, ϕ^2 is plotted against x, and the probability distributions in these states are seen to be markedly different from those expected classically. According to classical mechanics the particle is most likely to be found at the extremities of its motion, and for given energy these boundaries are quite sharp: in quantum mechanics there is clearly some possibility of penetration into the classically forbidden regions. As n increases, however, the quantum mechanical picture rapidly begins to approach that obtained from classical mechanics (see, for example, Pauling and Wilson (1935) p. 76).

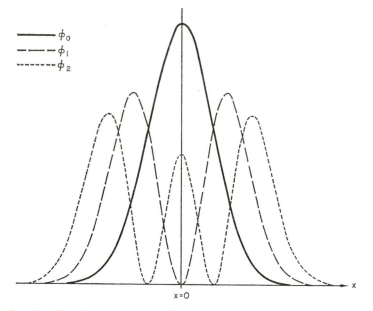

FIG. 2.4. The first three probability functions of the linear harmonic oscillator. The curves show $|\phi_n|^2$ for $n = 0, 1, 2$.

2.3. The hydrogen atom. Atomic units

The hydrogen atom will be treated fully elsewhere (Vol. 2), but it is useful to give here a simple treatment of the ground state. The proton, nearly 2000 times heavier than an electron, is considered to be at rest, merely defining the potential field in which the electron moves. The justification for this assumption is considered in Vol. 2, Section 2.1. We take the proton as the origin of coordinates and consider the one-particle Schrödinger equation for the electron. The classical potential energy expression is $V = -e^2/\kappa_0 r$, where r is the distance between the electron and the proton and $\kappa_0 = 4\pi\varepsilon_0$ (ε_0 = permittivity of free space).‡ Thus

$$H\phi = -\frac{\hbar^2}{2m}\nabla^2\phi - \frac{e^2}{\kappa_0 r}\phi = E\phi. \tag{2.20}$$

In view of the exponential behaviour of wave functions in a region of constant (or slowly varying) potential it would be reasonable to expect,

‡In this book we use SI units: κ_0 is replaced by unity in mixed Gaussian units.

for large r, solutions containing an exponential factor; in the simplest case we might even consider a "trial" function of the form

$$\phi(r) = A \exp(-br), \qquad (2.21)$$

where A and b are constants, and try to choose b so as to satisfy the equation (2.20). This procedure may or may not be successful, but may easily be tested. By direct differentiation, we obtain (retaining for the moment Cartesian coordinates, with $r = (x^2+y^2+z^2)^{1/2}$)

$$\frac{\partial^2 \phi}{\partial x^2} = Ab\left(-\frac{1}{r}+\frac{x^2}{r^3}+b\,\frac{x^2}{r^2}\right)\exp(-br)$$

and thus, adding the similar terms $\partial^2\phi/\partial y^2$ and $\partial^2\phi/\partial z^2$,

$$\nabla^2\phi = Ab\left(b-\frac{2}{r}\right)\exp(-br).$$

Putting this into (2.20), we require

$$\left[\frac{\hbar^2}{2m}b\left(b-\frac{2}{r}\right)+\frac{e^2}{\kappa_0 r}+E\right]A\exp(-br) = 0.$$

To make the expression in square brackets vanish, we must equate to zero separately the constant terms, giving

$$(\hbar^2/2m)b^2+E = 0,$$

and the terms in $1/r$, which give

$$(\hbar^2/m)b - e^2/\kappa_0 = 0.$$

Thus, when the constant is chosen as

$$b = me^2/\hbar^2\kappa_0 \qquad (2.22)$$

the wave function (2.21) satisfies the Schrödinger equation and the energy eigenvalue E is given by

$$E = -(\hbar^2/2m)b^2 = -(me^4/2\hbar^2\kappa_0^2). \qquad (2.23)$$

The remaining constant A is easily determined as

$$A = b^{3/2}/\sqrt{\pi} \qquad (2.24)$$

by use of the normalization condition.

The wave function (Fig. 2.5a) has no node and is therefore expected to represent the ground state. Further confirmation is obtained by inserting numerical values of m, e, κ_0 and $\hbar$ in (2.23): the result is

$$E = -2{\cdot}1796\times 10^{-18}\mathrm{J} = -13{\cdot}605\ \mathrm{eV}$$

and agrees fairly closely with the observed ionization potential of

13·595 eV‡—which is simply the energy required to remove the electron to infinity (the conventional zero of energy) from the hydrogen atom in its ground state.

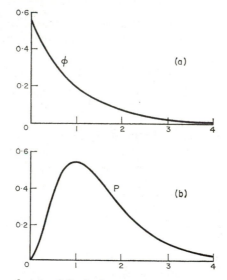

FIG. 2.5. Ground state of the hydrogen atom: (a) wave function ϕ as a function of radial distance r, (b) radial probability density P. Atomic units are used throughout.

A measure of the extension of the electron probability density in the ground state is given by $1/b$, since when r takes this value ϕ has fallen to $1/2·303$ times its value at the nucleus. In fact

$$b^{-1} = \kappa_0 \hbar^2 / m e^2 = 0·529172 \times 10^{-10} \text{ m}.$$

This is the "Bohr radius" and is usually denoted by a_0, its value being the radius of the first orbit in Bohr's semi-classical theory. To obtain the quantum mechanical interpretation of a_0, we remember that $|\phi|^2$ is a probability *per unit volume*, and the probability of finding the electron at a given distance, from the nucleus, say between limits r and $r+dr$, may thus be written $P(r)dr = 4\pi r^2 dr \times |\phi|^2$ since $|\phi|^2$ has the same value over the spherical shell of volume $4\pi r^2 dr$. The "radial probability density" P (Fig. 2.5b) therefore vanishes at the origin and has its maximum at a distance determined by setting $dP/dr = 0$: this condition at once yields $r = a_0$, showing that the Bohr radius actually indicates the most probable distance of the electron from the nucleus

---

‡The origin of the small difference of 0·01 eV will be discussed later: it arises mainly from the assumption of a fixed nucleus.

in the ground state. The wave function ϕ for an electron in an atom is usually referred to as an "atomic orbital"; it is commonly represented pictorially (Fig. 2.6) by means of a "contour map" showing surfaces on which ϕ is constant, or by indicating a surface (in this case a sphere) within which ϕ is largest and outside which the probability of finding the electron is small. The hydrogen-like atomic orbitals are fully discussed in Vol. 2.

Atomic units

In discussing atoms and molecules, it is often convenient to use "atomic units". In this system, the electronic mass m, the absolute value of the electron charge e, and the rationalized Planck constant

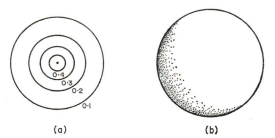

(a)   (b)

Fig. 2.6. Representations of hydrogen atomic orbitals: (a) contour map showing values of ϕ on spheres centred on the nucleus, (b) schematic indication of bounding contour (sphere) within which there is a high probability (e.g. 90%) of finding the electron.

$\hbar(= h/2\pi)$, are used as units of mass, charge and action, respectively. The units of energy, length, time, velocity are then completely determined and all equations take a simpler form, the variables then being numerical measures *in atomic units* of the quantities they represent. Thus, the energy and electronic wave function of the hydrogen atom in its ground state become

$$E = -\tfrac{1}{2} \text{ Hartree}, \quad \phi(r) = \pi^{-1/2} \exp{(-r)}$$

—the "Hartree" being fairly widely used as the name of the energy unit, though not regarded as a primary unit. Great care should be taken in using atomic units. For example, r in e^{-r} must be a pure number, for dimensional consistency, and the actual radial distance is thus ra_0. Also $|\phi(r)|^2$ is a *density* and $\phi(r)$ therefore implicitly contains a factor $a_0^{-3/2}$, although we have put $a_0 = 1$. A detailed discussion of units and dimensions is not required at present and is deferred until Vol. 2.

The energy is also frequently expressed in various "practical" units, such as electron volts, kcal mol^{-1},§ wave numbers (cm^{-1}),‖ etc. The relations among these units are indicated in Table 2.1.

TABLE 2.1
Conversion table for energy units

	Hartree	J	eV
1 Hartree	1	$4 \cdot 35944 \times 10^{-18}$	$27 \cdot 2108$
1 J	$2 \cdot 29364 \times 10^{17}$	1	$6 \cdot 24118 \times 10^{18}$
1 eV	$3 \cdot 67501 \times 10^{-2}$	$1 \cdot 60210 \times 10^{-19}$	1
1 kcal (mol^{-1})‡	$1 \cdot 59347 \times 10^{-3}$	$6 \cdot 94664 \times 10^{-21}$	$4 \cdot 33596 \times 10^{-2}$
1 cm^{-1}	$4 \cdot 55635 \times 10^{-6}$	$1 \cdot 98631 \times 10^{-23}$	$1 \cdot 23982 \times 10^{-4}$
1 MHz	$1 \cdot 51981 \times 10^{-10}$	$6 \cdot 62552 \times 10^{-28}$	$4 \cdot 13552 \times 10^{-9}$

	kcal (mol^{-1})‡	cm^{-1}	MHz
1 Hartree	$6 \cdot 27506 \times 10^2$	$2 \cdot 19474 \times 10^5$	$6 \cdot 57969 \times 10^9$
1 J	$1 \cdot 43917 \times 10^{20}$	$5 \cdot 03394 \times 10^{22}$	$1 \cdot 50930 \times 10^{27}$
1 eV	$23 \cdot 0609$	$8 \cdot 06569 \times 10^3$	$2 \cdot 41804 \times 10^8$
1 kcal (mol^{-1})‡	1	$3 \cdot 49725 \times 10^2$	$1 \cdot 04845 \times 10^7$
1 cm^{-1}	$2 \cdot 85914 \times 10^{-3}$	1	$2 \cdot 99794 \times 10^4$
1 MHz	$9 \cdot 53690 \times 10^{-8}$	$3 \cdot 33559 \times 10^{-5}$	1

Notes. The table indicates *equivalences*, not equalities. Thus 1 eV is "equivalent" to $23 \cdot 0609$ kcal(mol^{-1}) in so far as the latter is the energy of L systems (L = Avogadro number = $6 \cdot 02252 \times 10^{23}$) expressed in thermal units. Similarly, wave numbers and frequencies are related to corresponding energies by $E = h\nu$.

‡The thermochemical calorie is defined as 1 cal = $4 \cdot 184$ J.

2.4. The free particle

Although the particle moving in free space was considered in some detail in Chapter 1, one or two further points need attention. We choose the potential energy zero so that $V = 0$ everywhere and the Schrödinger equation becomes

$$\mathsf{H}\phi = -\frac{\hbar^2}{2m}\left(\frac{\partial^2\phi}{\partial x^2} + \frac{\partial^2\phi}{\partial y^2} + \frac{\partial^2\phi}{\partial z^2}\right) = E\phi. \tag{2.25}$$

§The energy in thermal units of L systems (L being the Avogadro number) instead of one: this is the unit used in thermochemistry.

‖An energy E may be associated with a *frequency* ν by the Planck relation $E = h\nu$, and hence with a wave number $(1/\lambda)$ by $E = hc(1/\lambda)$. Such units are employed in spectroscopy.

This has plane-wave solutions and by choosing the direction of propagation as the x-axis the dependence on y and z vanishes, permitting reduction to the one-dimensional equation

$$\frac{\partial^2 \phi}{\partial x^2} = -k^2 \phi \qquad (k^2 = 2mE/\hbar^2)$$

which has solutions (taking k positive) of the form‡

$$\phi = A \exp(ikx) + B \exp(-ikx), \quad E = (\hbar^2/2m)k^2. \qquad (2.26)$$

More generally, the equation may be separated (pp. 23–4 and Appendix 2) and the general solution is a product of three factors X, Y, Z, of similar form but referring to variables x, y, z, respectively.

In contrast with the other examples considered so far, the eigenvalues may here lie anywhere in the continuum of positive E values, for no restrictions have been placed on k; all positive energies are allowed. A second important difference is apparent when we try to normalize ϕ, for the integral of $|\phi|^2$ over all space is no longer finite for any non-zero values of A and B.

The reason for the normalization difficulty is in this case purely mathematical, for a particle known to be in any finite region of space at any given time may be described by a *wave packet*, formed by combining solutions of the type just found but with slightly different k values, and the *packet* is essentially localized and therefore normalizable although the basic wave trains are not. There are two main methods of circumventing this difficulty:

(i) We relax the requirement that $|\phi|^2$ be integrable, bearing in mind the fact that a normalized wave packet may always be built up from functions of almost identical wave numbers (so that it is necessary to consider the solution for only one value of k). In this case any mathematically convenient normalization may be used, and it is useful (for reasons which will appear presently) to choose the solutions

$$\phi_k(x) = (2\pi\hbar)^{-1/2} \exp(ikx), \quad E_k = (\hbar^2/2m)k^2 \qquad (2.27)$$

where the wave number k is allowed to take both positive and negative values in order to admit both possibilities in (2.26). The general three-dimensional solutions are correspondingly

$$\phi_k(\mathbf{r}) = (2\pi\hbar)^{-3/2} \exp(i\mathbf{k} \cdot \mathbf{r}),$$

$$E_k = (\hbar^2/2m)(k_x^2 + k_y^2 + k_z^2), \qquad (2.28)$$

---

‡Here it is most convenient to use $\exp(\pm ikx)$ instead of the equivalent real solutions $\sin kx$, $\cos kx$.

where **k** is a *wave vector* with components k_x, k_y, k_z which take positive or negative values, and $\mathbf{k} \cdot \mathbf{r} = xk_x + yk_y + zk_z$.

(ii) We imagine the particle to be confined within an arbitrarily large box, and normalize the wave function within the box. As the box is allowed to become indefinitely large the energy levels become quasi-continuous (cf. p. 26): for then $E = (\hbar^2/2m)k^2$ where $k = n\pi/L$ and n is a positive integer, and for L indefinitely large, k may be chosen as near as desired to any given positive number. There is a difference, however, in that the boundary conditions used previously rule out complex solutions of the form (2.27), only the sin (kx) combination being allowed in (2.26).

In order to obtain free-particle eigenfunctions with the box normalization we employ *periodic* boundary conditions of the form (2.10) and easily verify that solutions of the form $\phi(x) = C \exp(ikx)$ are then admitted, with $k = 2\pi n/L$ and n taking all integral values $(0, \pm 1, \pm 2, \ldots)$. Both terms in the general solution (2.26) are then covered. The solutions with "box normalization" are then

$$\phi_k(x) = \frac{1}{\sqrt{L}} \exp(ikx), \quad E_k = (\hbar^2/2m)k^2 \tag{2.29}$$

in the one-dimensional case, and

$$\phi_k(\mathbf{r}) = \frac{1}{\sqrt{L^3}} \exp(i\mathbf{k} \cdot \mathbf{r}), \quad E_k = (\hbar^2/2m)k^2 \tag{2.30}$$

in three dimensions. The allowed wave numbers are of the form

$$k_x, k_y, k_z = (2\pi/L) \times (0, \pm 1, \pm 2, \ldots) \tag{2.31}$$

and therefore become quasi-continuous for a very large box.

The second of the two methods of normalization is widely used in solid state theory and has a simple physical interpretation: the box may be regarded as a "fundamental volume" and the boundary conditions simply ensure that what happens in this volume is repeated indefinitely in all the similar volume elements formed by translation of the box along the three directions in space. The normalization implies that at the moment of observation the particle is supposed to be within the chosen fundamental volume: the corresponding $|\phi|^2$ is therefore a *conditional* probability, presupposing that a particle is present in the region under observation. The fact that the particle is "free", and is not reflected back at the boundaries of the hypothetical box, should be indicated by the existence of a non-zero probability current as defined in (1.30). For a particle with wave function $Ce^{i\mathbf{k}\cdot\mathbf{r}}$ the x component of the current density is

$$J_x = \frac{\hbar}{2im}\,|C|^2\left(e^{-ik_x x}\,ik_x e^{ik_x x} - e^{ik_x x}(-ik_x)e^{-ik_x x}\right) = |C|^2\frac{\hbar}{m}\,k_x \qquad (2.32)$$

and is uniform throughout the box. If we imagine a large number of completely independent particles, each behaving as it would with wave function $\phi_k(x)$, $|C|^2$ is the expected *number* of particles per unit volume, and J_x is the expected number per second crossing unit surface normal to the x-axis: classically this would be $J_x =$ (number/unit volume) × (x-component of velocity) and comparison with (2.32) shows that $\hbar k_x/m$ corresponds to the x-component of the particle velocity. In other words, k_x, k_y, k_z represent the three momentum components of the particle in units of $\hbar$. This interpretation will be confirmed more directly in a later section.

2.5. One-dimensional step potential with a finite potential height

Finally, we consider a free particle meeting a potential barrier of finite height. Let us choose the position of the barrier as the origin and the potential as

$$V(x) = 0 \quad (x < 0); \quad V(x) = V_0 > 0 \quad (x > 0).$$

This potential is illustrated in Fig. 2.7 and the wave functions are determined by

$$\mathsf{H}\phi = -\frac{\hbar^2}{2m}\frac{\partial^2\phi}{\partial x^2} + V(x)\phi = E\phi.$$

FIG. 2.7. One-dimensional potential barrier, showing typical wave function for a particle incident from the left with energy less than V_0. Reflection produces a standing wave.

We distinguish three cases, first $E < 0$, second $V_0 > E > 0$, third $E > V_0$ and discuss each in turn using the appropriate general solutions (p. 24):

(i) $E < 0$.

The solution is exponential in both regions. It soon appears, however, that the boundary conditions cannot be satisfied by any non-zero values of the constants. No acceptable wave function therefore exists, and, as in classical mechanics, there is no corresponding state of motion with $E < 0$.

(ii) $0 < E < V_0$.

It is clear that ϕ must be of complex exponential form in the left-hand region of Fig. 2.7, but must be a decaying exponential in the barrier region: let us take

$$\phi(x) = e^{ikx} + A e^{-ikx}, \quad k = \sqrt{(2mE/\hbar^2)} \qquad (x < 0),$$
$$\phi(x) = B e^{-k'x}, \qquad k' = \sqrt{\{2m(V-E)/\hbar^2\}} \quad (x > 0). \quad (2.33)$$

Normalization has been disregarded in view of the discussion in Section 2.4. The boundary conditions at $x = 0$ are easily found to be

$$1 + A = B,$$
$$ik - Aik = -Bk',$$

from which we obtain

$$A = -\frac{k' + ik}{k' - ik}, \quad B = \frac{-2ik}{k' - ik}. \qquad (2.34)$$

To interpret the solution we compute the probability current density in each region. For $x < 0$ we obtain

$$J_x = \frac{\hbar}{2im} \left\{ (e^{-ikx} + A^* e^{ikx})(ik e^{ikx} - Aik e^{ikx}) \right.$$
$$\left. - (e^{ikx} + A e^{-ikx})(-ik e^{-ikx} + A^* ik e^{ikx}) \right\}$$

which reduces to

$$J_x = \frac{\hbar k}{m}(1 - |A|^2). \qquad (2.35)$$

From (2.34), however, it follows that A is a unimodular complex number; J_x therefore vanishes. In the absence of the barrier ($A = B = 0$), there would be a probability current $\hbar k/m$ along the

x-direction;‡ and the fact that insertion of the barrier reduces this current to zero means that a particle with $E < V_0$ is sure to be reflected. The solution may in fact be written in real form and describes a standing wave on the left of the barrier. There is clearly a non-zero probability that the particle will penetrate some distance into the barrier; but it is easily verified that, for $V_0 \to \infty$, $\phi(x) \to \sin kx$ ($x < 0$), $\to 0$ ($x > 0$). Thus for sufficiently large V_0 the barrier forms a strictly impenetrable wall and the boundary condition imposed in Section 2.1 is rigorously justified.

(iii) $E > V_0$.

The solution must be of complex exponential form in both regions, a single term $e^{ik'x}$ being appropriate for $x > 0$ since the term $e^{-ik'x}$ would describe a particle incident from the *right* of the barrier, a situation not considered. The values of k (for the region $x < 0$) and k' (for $x > 0$) are evidently

$$k = \sqrt{(2mE/\hbar^2)}, \quad k' = \sqrt{\{2m(E-V_0)/\hbar^2\}},$$

and we find easily the current density expressions

$$J_x = \frac{\hbar k}{m}(1 - |A|^2) \quad (x < 0) \tag{2.35a}$$

$$J_x = \frac{\hbar k'}{m}|B|^2 \quad (x > 0), \tag{2.35b}$$

where

$$A = \frac{k-k'}{k+k'}, \quad B = \frac{2k}{k+k'}. \tag{2.36}$$

The statistical interpretation is clearest if we imagine a large number of independent particles, each described by the wave function ϕ. In this case the number per unit volume is proportional to $|\phi|^2$ and the number flowing across unit area perpendicular to the x-axis is proportional to J_x. We then have

Number crossing unit area/sec
   (i) in absence of the barrier $\propto \hbar k/m$,
   (ii) to left of the barrier $\propto \hbar k/m \times (1 - |A|^2)$,
   (iii) in the region beyond the barrier $\propto \hbar k/m \times (k'/k)|B|^2$.

It follows that $|A|^2$ represents the fractional number of particles turned back from the barrier (thereby reducing the *net* current) and is therefore a *reflection coefficient*, while $(k'/k)|B|^2$ is the fractional number penetrating the barrier, or in other words a *transmission coefficient*. The sum of the two coefficients is of course unity, any incident particle being either transmitted or reflected.

It should be noted that the above results have been obtained using the *stationary state* eigenvalue equation, which may appear to be somewhat artificial. We remark that exactly the same results may be obtained by constructing an incident wave packet, using wave numbers in the immediate vicinity of k, and following the time development of the packet as it meets the barrier. The wave packet is simply a superposition of the stationary state solutions, of which we have obtained a typical one. The wave packet treatment of scattering processes is fully discussed in another volume of this series.

REFERENCE

PAULING, L. and WILSON, E. B. (1935) *Introduction to Quantum Mechanics*, McGraw Hill, New York.

CHAPTER 3

MATHEMATICAL DIGRESSION

3.1. Preliminaries. Operators and eigenvalue equations

In Chapter 2 we set up and solved Schrödinger's equation for certain simple one-particle systems. Most of the systems of interest in physics and chemistry, however, are vastly more complex, and many mathematical generalizations are necessary before progress can be made; some of these generalizations relate to concepts and principles, with which this volume is mainly concerned; others relate to the development of methods for obtaining approximate solutions and are studied more fully in Vol. 2. In this Chapter we review the mathematical techniques which are basic to the general formulation of quantum mechanics.

We start from the differential equations encountered in Chapter 2. These are all *eigenvalue equations*; each one contains some numerical parameter, λ say, such that solutions of a given class (p. 18) arise only when λ takes specific values—the *eigenvalues*. The set of eigenvalues— the *spectrum*—may be discrete (as in the examples of Sections 2.1–2.3) or continuous (as in Sections 2.4–2.5); but in all cases we meet equations of the general form

$$a(x) \frac{d^2\psi}{dx^2} + b(x) \frac{d\psi}{dx} + c(x)\psi = \lambda f(x)\psi, \qquad (3.1)$$

where a, b, c, f are real functions of x, while $\psi(x)$ may be the wave function itself in a one-dimensional problem, or one of the factors when the wave function is written as a product in separating the variables.

The left-hand side of (3.1) may be regarded as a prescription for obtaining from ψ a new function; in words it means "multiply ψ by the function $c(x)$, to this add the result of differentiating ψ with respect to x and then multiplying by $b(x)$, and finally add the result of differentiating ψ *twice* and multiplying by $a(x)$". Each term is the result of a certain *operation* on ψ, and it is customary to speak of $b(x)(d/dx)$, for example, as the *operator* involved in obtaining from the operand ψ the new function $b(x)d\psi/dx$. It is then possible to recognize properties of the operators which do not depend on the particular form of the function on which they operate. For example,

$$b(x) \frac{d}{dx} (\psi_1 + \psi_2) = b(x) \frac{d}{dx} \psi_1 + b(x) \frac{d}{dx} \psi_2$$

for *any* pair ψ_1, and ψ_2, of continuous functions of x. If we abbreviate the differential operator to the single symbol D this result, and a similar one obtained by operating on a multiple of ψ, may be written

$$D(\psi_1 + \psi_2) = D\psi_1 + D\psi_2,$$
$$D(c\psi) = c\,D\psi \tag{3.2}$$

for any complex number c. These are not trivial properties (they are not possessed, for example, by the operator, S say, which squares the operand); they serve to define *linear operators*. Nearly all the operators we encounter in quantum mechanics are linear.

It is also convenient to define the sum and the product of operators, in such a way that, for example, the left-hand side of (3.1) may be denoted by D where

$$D = a(x)\,\frac{d^2}{dx^2} + b(x)\,\frac{d}{dx} + c(x)$$

and the implication is that when D operates on *any* function ψ it produces the function on the left in equation (3.1). The *sum* of two operators A and B is thus defined so that

$$(A + B)\psi = A\psi + B\psi. \tag{3.3}$$

In other words, $(A + B)$ operating on any ψ *means* "operate with A and B separately and add the results". Similarly, the product (AB) is defined by

$$(AB)\psi = A(B\psi) \tag{3.4}$$

and (AB) operating on any ψ means "operate first with B (nearest to ψ) and then operate on the result with A". If A and B stand for the operations of multiplying ψ by $a(x)$ and $b(x)$ respectively, it is clear that $AB = BA$ (the order of multiplication makes no difference); the operators are then said to *commute*. But when differential operators are admitted it is easily verified that the order is important and in general $AB \neq BA$. It is also clear that the notion of *equality* of two operators needs definition; for generality we must define

$$A = B \quad \textit{if and only if} \quad A\psi = B\psi \ (\text{all } \psi), \tag{3.5}$$

where "all ψ" means for all functions of the class considered in defining the operators.‡ The most trivial operation is multiplication by a

---

‡Normally it will be assumed that the functions of importance in quantum mechanics belong to class Q (p. 18). It is important to note that the operators are fully defined only with reference to some specified class (e.g. functions for which the derivatives *exist*), although points of mathematical rigour will usually be ignored.

constant, c. The above definitions simply ensure that any collection of sums, products, and multiples of operators may be manipulated exactly as in elementary algebra *provided the order of the operators is always retained*. In this connection we note that the usual shorthand for powers is also taken over. For example,

$$(a\mathsf{A} + b\mathsf{B})^2 = (a\mathsf{A} + b\mathsf{B})(a\mathsf{A} + b\mathsf{B}) = a^2\mathsf{A}^2 + ab(\mathsf{AB} + \mathsf{BA}) + b^2\mathsf{B}^2$$

where the middle term reduces to $2ab\mathsf{AB}$ only if the operators A and B commute.

Eigenvalue equations. Self-adjoint form

We shall study some general properties of eigenvalue equations, writing (3.1) in the form

$$\mathsf{D}\psi = \lambda f \psi \tag{3.6}$$

where the differential operator D now denotes (a, b, c, still functions of x)

$$\mathsf{D} = a\frac{d^2}{dx^2} + b\frac{d}{dx} + c. \tag{3.7}$$

The general theory of linear second order equations is usually developed, however, not for (3.6) but for the "Sturm–Liouville equation"

$$\mathsf{L}\psi = \lambda w \psi, \tag{3.8}$$

where L is a "self-adjoint" operator with the rather specific form

$$\mathsf{L} = \frac{d}{dx}\left(p\frac{d}{dx}\right) + q = p\frac{d^2}{dx^2} + \frac{dp}{dx}\frac{d}{dx} + q \tag{3.9}$$

in which $p(x)$, $q(x)$ are real functions of x, while the $w(x)$ is everywhere real and positive. Much is known about Sturm–Liouville equations (Courant and Hilbert, 1953; Kemble, 1937) and we therefore note first that the general equation (3.6) may always be reduced to the form (3.8) in which L is self-adjoint. To do this we multiply (3.7) by a function $u(x)$ and try to choose $u(x)$ so that

$$u\mathsf{D} = au\frac{d^2}{dx^2} + bu\frac{d}{dx} + cu = \mathsf{L}$$

$$uf = w. \tag{3.10}$$

On comparing $u\mathsf{D}$ with (3.9) we must choose u so that

$$bu = \frac{d}{dx}(au). \tag{3.11}$$

But this is a simple first-order equation for the unknown function $u(x)$ and can always be integrated to give the desired result. *Any* eigenvalue equation defined by (3.6) and (3.7) may thus be written in Sturm–Liouville form, and the results of Sturm–Liouville theory are of rather general applicability. We continue to assume that the function w, whose significance becomes apparent presently, is everywhere positive; this assumption covers all the most important equations.

EXAMPLE. *Hermite's equation.* This equation, namely

$$\frac{d^2y}{dx^2} - 2x\frac{dy}{dx} + 2\beta y = 0,$$

which arose in the form (2.14) from the harmonic oscillator problem, is not in Sturm–Liouville form. On multiplying through by u, we can make the differential operator self-adjoint provided we choose u so that $-2xu = (du/dx)$. A solution is $u = e^{-x^2}$ and the equation can therefore be written in the standard form (3.8) with

$$\mathsf{L} = e^{-x^2}\frac{d^2}{dx^2} - 2xe^{-x^2}\frac{d}{dx}, \quad w = e^{-x^2}, \quad \lambda = -2\beta.$$

The weight factor w, introduced in this way, is particularly significant, as will be seen presently.

We now show that the property of self-adjointness, together with certain rather general boundary conditions, implies that L has a very simple symmetry property. If we are concerned with solutions defined in the interval (a, b) and u, v are *any* two functions of the given class, the symmetry property which results is:

$$\int_a^b u(\mathsf{L}v)\,dx = \int_a^b (\mathsf{L}u)v\,dx. \tag{3.12}$$

To show this, we write L in the first form of (3.9) and integrate by parts:

$$\int_a^b v(\mathsf{L}u)\,dx = \int_a^b v\frac{d}{dx}\left(p\frac{du}{dx}\right)dx - \int_a^b vqu\,dx$$

$$= \left[v\left(p\frac{du}{dx}\right)\right]_a^b - \int_a^b \left(p\frac{du}{dx}\right)\frac{dv}{dx}dx - \int_a^b vqu\,dx.$$

Now the boundary conditions appropriate in quantum mechanics are invariably

$$\left(vp\frac{du}{dx}\right)_{x=a} = \left(vp\frac{du}{dx}\right)_{x=b}. \tag{3.13}$$

Both quantities vanish when the boundaries are at infinity, owing to the condition of quadratic integrability; and the equality is also satisfied by the periodic boundary conditions of Section 2.4. On adopting this condition the first term in the expression for the transformed integral vanishes; and on performing another integration by parts we obtain the desired result.

To anticipate the appearance of *complex* functions and operators we state a slightly more general symmetry property which includes (3.12) as a special case. We normally assume quadratically integrable functions (class Q) and when

$$\int_a^b u^*(\mathsf{L}v)\,dx = \int_a^b (\mathsf{L}u)^*v\,dx \qquad (3.14)$$

for any two functions of the class, the operator L (self-adjoint in a wider sense) is usually said to be *Hermitian*. We return to Hermitian operators in Section 3.7.

Properties of eigenfunctions

The property of self-adjointness of an operator (or more generally of Hermitian symmetry) is reflected in certain very general properties of its eigenfunctions. We suppose that ϕ_i and ϕ_j are any two solutions of (3.8) and assume that L has Hermitian symmetry. Then by definition

$$\mathsf{L}\phi_i = \lambda_i w\phi_i,$$

$$\mathsf{L}\phi_j = \lambda_j w\phi_j.$$

We multiply the first equation by ϕ_j^*, and the complex conjugate of the second by ϕ_i, and then integrate each over the range (a, b) to obtain (remembering w is assumed real)

$$\int_a^b \phi_j^*\mathsf{L}\phi_i\,dx = \lambda_i \int_a^b w\phi_j^*\phi_i\,dx,$$

$$\int_a^b (\mathsf{L}\phi_j)^*\phi_i\,dx = \lambda_j^* \int_a^b w\phi_j^*\phi_i\,dx.$$

But by (3.14) the left-hand sides are equal and hence, by subtraction,

$$(\lambda_i - \lambda_j^*)\int_a^b \phi_j^*\phi_i w\,dx = 0. \qquad (3.15)$$

Now if we choose $\phi_i = \phi_j$ the integrand becomes $|\phi_i|^2 w$ and must be everywhere positive: the conclusion is that $\lambda_i = \lambda_i^*$. In other words the *eigenvalues of a Hermitian operator are real*.

We pass to the case $i \neq j$ and observe that if $\lambda_i \neq \lambda_j$ then

$$\int_a^b \phi_i^*\phi_j w\,dx = 0 \qquad (\lambda_i \neq \lambda_j). \qquad (3.16)$$

Functions with the property (3.16) are said to be "orthogonal with weight factor w in the interval (a, b)". Thus *eigenfunctions of* (3.8) *with different eigenvalues are mutually orthogonal in the sense* (3.16).

In most of the examples discussed so far the weight factor w has been unity and the eigenfunctions have been simply "orthogonal" without qualification. In the one-dimensional box, for example (Section 2.1), the eigenfunctions have the well-known property

$$\int_0^L \sin(n\pi x/L) \sin(n'\pi x/L)\, dx = 0 \quad (n' \neq n)$$

conforming to (3.16) with $w = 1$. The harmonic oscillator (Section 2.2) provides the following more interesting example.

EXAMPLE. *Hermite's equation.* The solutions of class Q in the interval $(-\infty, +\infty)$, denoted by $H_n(x)$, arise when $\beta = n$, an integer. Since the weight factor occurring when the equation is written in Sturm–Liouville form (p. 44) is $w = e^{-x^2}$ the orthogonality property of different eigenfunctions is

$$\int_{-\infty}^{+\infty} H_n(x)H_{n'}(x)e^{-x^2}\, dx = 0 \quad (n' \neq n).$$

We note also that the functions $\phi_n(x) = H_n(x)e^{-x^2/2}$ are orthogonal in the simple sense

$$\int_{-\infty}^{+\infty} \phi_n(x)\, \phi_{n'}(x)\, dx = 0 \quad (n' \neq n)$$

and notice that (apart from a change of variable $x \rightarrow x\sqrt{\alpha}$) these are solutions of the original quantum mechanical problem (2.12). The argument leading to (3.16) could in fact have been applied directly to (2.12), which is already in Sturm–Liouville form with $w = 1$. Here we proceeded via Hermite's equation partly because this had already occurred in reducing (2.12) to a known standard form, and partly to illustrate orthogonality in the wider sense with $w \neq 1$.

Finally, we note from (3.15) that two different functions ϕ_i, ϕ_j do not necessarily satisfy (3.16) if λ_i and λ_j happen to be equal. In this case of degeneracy it is also true from the linearity of the operator L that any linear combination $\phi = a\phi_i + b\phi_j$ will also satisfy (3.8) with $\lambda = \lambda_i = \lambda_j$ (as may be verified by substitution). This means there is no need to discuss non-orthogonal solutions: for from ϕ_i and ϕ_j we can always construct two mixtures, ϕ_i' and ϕ_j', which are still solutions and at the same time are *orthogonal*. The simplest way of doing this is indicated in the following example.

EXAMPLE. *Orthogonalization.* Suppose ϕ_1 and ϕ_2 are two degenerate eigenfunctions, each satisfying $L\phi = \lambda w \phi$. Now replace the second function by

$\phi_2' = \phi_2 + c\phi_1$ and choose c so that ϕ_1 and ϕ_2' are orthogonal in the sense

$$\int_a^b \phi_1^*(\phi_2 + c\phi_1)w\,dx = 0.$$

This requires only that

$$c = -\int_a^b \phi_1^*\phi_2 w\,dx \bigg/ \int_a^b \phi_1^*\phi_1 w\,dx$$

and, since (by assumption $w > 0$) the denominator is non-zero, a solution certainly exists. Given a third degenerate solution say, ϕ_3, we could replace it by ϕ_3', orthogonal to both ϕ_1 and ϕ_2' (already orthogonal), by choosing $\phi_3' = \phi_3 + c_1\phi_1 + c_2\phi_2'$ and eliminating both coefficients from the *two* orthogonality conditions. In fact the procedure could be continued for any number of degenerate functions, to yield a new degenerate set of *orthogonal* functions. This method of orthogonalizing is the *Schmidt process*, which can be applied to *any* set of functions, degenerate or non-degenerate.

The conclusion from the last Example is that a set of eigenfunctions of a Sturm–Liouville type equation may without loss of generality be assumed orthogonal, irrespective of any degeneracies which may occur. If orthogonalization is necessary it may be achieved in an infinite variety of ways; the Schmidt process indicated in the above Example is simply one systematic construction. It is customary also to *normalize* each member of the set so that $\int \phi_n^*\phi_n w\,dx = 1$, and the set is then said to be *orthonormal*. The properties of normality and orthogonality are usually combined in the statement

$$\int_a^b \phi_i^*\phi_j w\,dx = \delta_{ij}, \qquad (3.17)$$

where $\delta_{ij} = 1\ (i = j)$, $= 0\ (i \neq j)$ is called the "Kronecker delta" and is an extremely useful symbol.

3.2. Eigenfunction expansions

We recall that an arbitrary function $f(\theta)$ can be expressed in the interval $-\pi \leqq \theta \leqq +\pi$ in terms of the functions

$$\sin n\theta, \quad \cos n\theta, \quad (n = 0, 1, 2, \ldots)$$

according to Fourier's theorem. If, for simplicity, we confine attention to functions defined in the region $0 \leqq \theta \leqq \pi$ and satisfying the boundary conditions $f(0) = f(\pi) = 0$ the cosine terms may be omitted and we get a very simple example of an *eigenfunction expansion*

$$f(\theta) = \sum_{n=1}^{\infty} c_n \phi_n(\theta).$$

The expansion functions

$$\phi_n(\theta) = \frac{1}{\sqrt{\pi}} \sin n\theta$$

are immediately recognized as eigenfunctions of the box problem of Section 2.1 in which $\pi x/L$ is replaced by the variable θ. It is easily verified that the (normalized) functions $\phi_n(\theta)$ form an orthogonal set, as required by Sturm–Liouville theory. In quantum mechanics, the standard method of obtaining approximate eigenfunctions of some complicated eigenvalue equation is to introduce a suitable orthonormal set $\{\phi_n\}$ and to build up a best approximation by combining a *finite* number of terms with properly chosen coefficients. It is useful, even at this point, to examine the nature of the approximation involved.

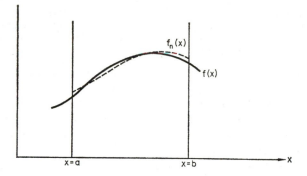

FIG. 3.1. Curve fitting. The function $f(x)$ is approximated in the interval (a, b) by $f_n(x)$.

Suppose we consider the expansion of $f(x)$, in the interval (a, b), in terms of a given set of functions $\{\phi_i(x)\}$ $(i = 1, 2, \ldots)$. First we use only the first n terms, writing

$$f(x) \simeq f_n(x) = \sum_{i=1}^{n} c_i \phi_i(x) \tag{3.18}$$

as an n-term approximation to $f(x)$. The problem then is to choose the coefficients to get a *best possible* representation of this form, and this may be done in a convenient way by making the mean square deviation (Fig. 3.1)

$$M_n = \int_a^b |f(x) - \sum_{i=1}^{n} c_i \phi_i(x)|^2 dx \tag{3.19}$$

as small as possible. This is a simple "curve-fitting" problem. It also introduces a new and important idea, that of *completeness*. If an arbitrary function $f(x)$ can be approached arbitrarily closely in the sense $M_n \to 0$ as $n \to \infty$, the set $\{\phi_i(x)\}$ is said to be *complete*. Because the definition involves integration it does not imply that the right- and left-hand sides of (3.18) should become equal at *every point*, in the limit, but rather that if any deviations occur they do so only at a *finite number of points* and in such a way as to give zero contribution to the integral. The function is said to be approximated "in the (quadratic) mean" and it should be noticed that the approximation is different in character from, say, a Taylor expansion, where the arbitrary function is approached arbitrarily closely at *all* points in the neighbourhood of a given point. We return later to the question of completeness.

The importance attached to deviations in different regions could be adjusted by introducing an arbitrary weight factor, replacing (3.19) by

$$M_n = \int_a^b \left| f(x) - \sum_{i-1}^n c_i \phi_i(x) \right|^2 w(x)\, dx \qquad (3.20)$$

and we study the minimization of this more general measure of the deviation. The best coefficients occur when‡ $\partial M/\partial c_i = 0$, $\partial M/\partial c_i^* = 0$ $(i = 1, 2, \ldots, n)$ and taking the second condition gives

$$\int_a^b \left(f(x) - \sum_{=1}^n c_j \phi_j(x) \right) \phi_i^*(x) w(x)\, dx = 0.$$

Thus for each function ϕ_i we require

$$\int_a^b \phi_i(x) f(x) w(x)\, dx = \sum_{j=1}^n c_j \int_a^b \phi_i^*(x) \phi_j(x) w(x)\, dx. \qquad (3.21)$$

An important practical consideration now arises. If a function $\phi_{n+1}(x)$ is added to the expansion, it would be most inconvenient to have to redetermine the coefficients of the first n functions: and there is then good reason to impose a "condition of finality", trying to choose the function system in such a way that the best values of the first n coefficients in an $(n+1)$ term approximant agree exactly with those of the n-term approximant. If we write down the counterpart of the last equation it will contain just one new term on the right-hand side.

---

‡Note that if F is a function of a complex number c it is permissible to treat c and c^* as independent variables. Thus, the real part x and the imaginary part y can vary independently, and an extremum of F is therefore given by the conditions

$$\frac{\partial F}{\partial x} = \frac{\partial F}{\partial c}\frac{\partial c}{\partial x} + \frac{\partial F}{\partial c^*}\frac{\partial c^*}{\partial x} = \frac{\partial F}{\partial c} + \frac{\partial F}{\partial c^*} = 0,$$

$$\frac{\partial F}{\partial y} = \frac{\partial F}{\partial c}\frac{\partial c}{\partial y} + \frac{\partial F}{\partial c^*}\frac{\partial c^*}{\partial y} = \frac{\partial F}{\partial c} - \frac{\partial F}{\partial c^*} = 0.$$

These conditions are clearly equivalent to $(\partial F/\partial c) = (\partial F/\partial c^*) = 0$. Thus it is formally possible to regard c_i and c_i^* as independent variables, even though knowledge of one completely determines the other.

Subtraction of the two equations then shows that the condition of finality is equivalent to the requirement

$$\int_a^b \phi_{n+1}^*(x)\phi(x)w(x)\,dx = 0 \quad (i = 1, 2, \ldots, n).$$

In other words, $\phi_{n+1}(x)$ must be orthogonal to $\phi_1(x), \ldots, \phi_n(x)$ with weight factor $w(x)$. It follows by induction that *every pair* of functions in the set must satisfy a similar orthogonality condition and, if (as may always be done) we normalize each function, the set will be *orthonormal*:

$$\int_a^b \phi_i^*(x)\phi_j(x)w(x)\,dx = \delta_{ij}. \tag{3.22}$$

The sets of eigenfunctions of differential equations, introduced in earlier sections, therefore seem to be excellently suited for approximation in the mean to arbitrary functions of the same class.

For orthonormal sets the expansion coefficients follow easily, for in (3.21) only one term remains on the right-hand side—that which contains c_i. The result is then

$$c_i = \int_a^b \phi_i^*(x)f(x)w(x)\,dx. \tag{3.23}$$

This determination of coefficients is final and unchanged as the approximation is improved by taking more and more terms.

We can now express the idea of completeness in a more precise form by calculating the quantity M_n and examining its behaviour as $n \to \infty$.

$$M_n = \int_a^b [\,|f|^2 - f^*f_n - f_n^*f + |f_n|^2\,]\,w\,dx$$

$$= \int_a^b |f|^2 w\,dx - \sum_{j=1}^n c_j \int_a^b f^*\phi_j w\,dx - \sum_{i=1}^n c_i^* \int_a^b \phi_i^* f w\,dx + \sum_{i,j}^n c_i^* c_j \int_a^b \phi_i^*\phi_j w\,dx$$

$$= \int_a^b |f|^2 w\,dx - 2\sum_{j=1}^n |c_j|^2 + \sum_{j=1}^n |c_j|^2$$

and since M_n is by definition essentially positive, this requires

$$\int_a^b |f|^2 w\,dx - \sum_{j=1}^n |c_j|^2 > 0. \tag{3.24}$$

This is "Bessel's inequality". Clearly, completeness requires that the quantity on the left tends to zero as $n \to \infty$: at any stage in the process it gives a convenient numerical measure of the failure of a *finite* approximation. In quantum mechanics we are usually concerned with functions of integrable square, normalized so that the first term in (3.24) has the value unity, and in this case completeness of the set requires that in the expansion of *any* normalized function of the given class

$$\sum_{j=1}^{n} |c_i|^2 \to 1$$

(from below), as $n \to \infty$. Important theorems concerning complete sets have been established in recent years, but for present purposes the completeness criterion implicit in Bessel's inequality will be adequate.

3.3. Generalization to many variables

Although the preceding discussion of eigenvalue equations and eigenfunction expansions arose from a study of differential equations of Sturm–Liouville type most of the arguments were, at least superficially, independent of the mathematical nature of the operator L and the function ψ. In quantum mechanics we deal with functions of *many* variables and note the following generalizations:

(i) *Orthonormality of eigenfunctions*

In introducing this idea we assumed only *one* independent variable x. We might, however, have worked equally well with Schrödinger's equation in unseparated form, namely $H\psi = E\psi$ where (for one particle) H and ψ involve *three* variables. Or we might have considered the most general case in which‡ $\Psi = \Psi(\mathbf{x})$ is a many-particle wave function, involving *many* variables which we symbolize collective by a *bold* letter $\mathbf{x}$. Provided H has the hermitian symmetry property analogous to (3.14), namely,

$$\int U^* H V \, d\mathbf{x} = \int (HU)^* V \, d\mathbf{x} \tag{3.25}$$

then the orthogonality of different solutions of the general equation

$$H\Psi = E\Psi \tag{3.26}$$

follows exactly as in the one-variable case (pp. 45–46). We note that the weight factor in (3.26) is unity (cf. (3.8)) and that the eigenfunctions may thus be assumed orthonormal in the simple sense

$$\int \Psi_i^* \Psi_j \, d\mathbf{x} = \delta_{ij}, \tag{3.27}$$

---

‡It is convenient in later chapters to use capital letters to denote general many-particle wave functions, reserving small letters for statements that are valid only for single-particle functions,

where $\int d\mathbf{x}$ now implies integration over the whole many-dimensional space. We shall find later that (provided Ψ vanishes at infinity or satisfies periodic boundary conditions analogous to those included in (3.13)) H is indeed self-adjoint in the generalized sense (3.25).

It is natural to enquire whether the generalization from one to several variables extends also to the eigenfunction expansion, and it is at once clear that the original arguments (pp. 48–51) remain valid: if

$$\Psi(\mathbf{x}) = \sum_i c_i \Phi_i(\mathbf{x}) \tag{3.28}$$

is an expansion of a function of some given class in terms of orthonormal functions Φ_i of the same class, then the optimum choice of coefficients is given by

$$c_i = \int \Phi_i{}^*(\mathbf{x})\Psi(\mathbf{x})d\mathbf{x} \tag{3.29}$$

exactly as in (3.23). Again we may discuss completeness in terms of the quantity M_n defined as in (3.20) with $w = 1$, arriving at the criterion $M_n \to 0$ for $n \to \infty$. This, of course, does not guarantee that complete sets exist in general; but many instances are known in which existence can be proved (e.g. wave functions of the three-dimensional box problem, discussed in Section 2.1) and this encourages the belief that— even for functions of several variables—eigenfunctions of an operator such as H normally provide complete sets. In quantum mechanics, we therefore simply accept completeness, in the general case of functions of many variables, as a postulate.

(ii) *Adjoint operators and Hermitian symmetry*

The concept of self-adjoint or Hermitian operators also needs extension. Let H be any operator, defined with respect to some complete set of functions $\{\Phi_i\}$ on which it operates, and let us associate with H the *adjoint operator* H† by the definition

$$\int \Phi^* H^\dagger \Phi \, d\mathbf{x} = \left(\int \Phi^* H\Phi \, d\mathbf{x} \right)^* \quad \text{(all } \Phi) \tag{3.30}$$

where "all Φ" means for arbitrary functions of the class defined by $\{\Phi_i\}$. It then follows easily‡ that an equivalent definition is

$$\int \Phi_i{}^* H^\dagger \Phi_j d\mathbf{x} = \int (H\Phi_i)^* \Phi_j d\mathbf{x}. \tag{3.31}$$

An operator is naturally called *self*-adjoint if H† = H and the definition, in the form (3.31), therefore yields

$$\int (H\Phi_i)^* \Phi_j d\mathbf{x} = \int \Phi_i{}^* H\Phi_j d\mathbf{x}. \tag{3.32}$$

Evidently this symmetrical form of (3.31), appropriate when H† = H,

---

‡Use Φ first as $\Phi_i + i\Phi_j$ and second as $\Phi_i - i\Phi_j$ and then subtract the resultant equations, observing that for any two functions $\int \Phi_j{}^* \Phi_i d\mathbf{x} = [\int \Phi_i{}^* \Phi_j d\mathbf{x}]^*$.

provides a powerful generalization of (3.14) which was formulated for functions of one variable with a particular type of differential operator.

The physical implication of Hermitian symmetry follows at once on taking for Φ in (3.30) any normalized eigenfunction Ψ_i of H with eigenvalue E_i: for then $\int \Psi_i * H \Psi_i dx = E_i \int \Psi_i * \Psi_i dx = E_i$ and (3.30) states that $E_i^* = E_i$. *The eigenvalues of a Hermitian operator are essentially real.* The association of Hermitian operators with dynamical quantities (e.g. the Hamiltonian operator with the energy) therefore seems specially appropriate, in so far as the measured values of observables are essentially real quantities.

Most of the operators we encounter in quantum mechanics are Hermitian, but some are not. It is therefore useful to note that *any* operator can be written in the form $C = A + iB$ where A and B are Hermitian, and that the adjoint is then obtained simply by reversing the sign of i:

$$C = A + iB, \quad C^\dagger = A - iB. \tag{3.33}$$

We meet such operators in the theory of angular momentum.

(iii) *Complete sets as product functions*

Finally, we note that when eigenfunctions appear in separated form the factors refer to independent complete sets, one for each variable. Thus in the three-dimensional box problem (Section 2.1) the eigenfunctions were of the form

$$\phi_{lmn}(x, y, z) = X_l(x) Y_m(y) Z_n(z)$$

where $\{X_l(x), l = 1, 2, \ldots\}$ is complete for functions of x in the range $(0, L)$ etc. In this way, a complete set in three variables is evidently expressed in terms of three complete sets each in a single variable. This observation may be stated as follows:

> If $\{u_i(x)\}$ and $\{v_j(y)\}$ are complete sets for functions of x in the interval (a, b), and of y in the interval (c, d), respectively, then the set of all possible products $\{u_i(x)v_j(y)\}$ is complete for functions $f(x, y)$ of both variables, defined in the same intervals. $\tag{3.34}$

This simply means that an arbitrary function $f(x, y)$ may be expanded in the form

$$f(x, y) = \sum_{i,j} c_{ij} u_i(x) v_j(x) \tag{3.35}$$

and that as we take more and more product functions we shall get convergence in the mean over the whole region bounded by $x = a$,

$x = b$ and $y = c$, $y = d$. As usual, it is difficult to give general and mathematically rigorous proofs of such assertions. The result (3.35) becomes highly plausible, however, if we first expand $f(x, y)$ using the assumed completeness of $\{u_i(x)\}$ in the form

$$f(x, y) = \sum_i c_i u_i(x)$$

for any fixed value of y; and then note that the coefficients must depend on y only and may consequently be expanded in the form

$$c_i(y) = \sum_j c_{ij} v_j(y).$$

On inserting this expression for the coefficients we then obtain (3.35). This device is used freely in quantum mechanics for building complete sets in many variables from complete sets in fewer variables. In such applications the intervals involved are frequently infinite and the validity of the procedure is normally established by usage rather than by rigorous analysis.

3.4.   Linear vector spaces. Basic ideas‡

Suppose we have some class of functions, for example the set of all single-valued continuous quadratically integrable functions of a variable x defined in an interval (a, b). If ψ_1 and ψ_2 are any two members of this class, their sum possesses the same basic properties and is therefore another member of the same class. So is any multiple $c\psi$, c being an arbitrary number, and hence $c_1\psi_1 + c_2\psi_2$ is also a member of the class. Also§ $c\psi = \psi$ only for $c = 1$ and $c\psi = 0$ only for $c = 0$. These properties are essentially those which define a *vector space*: in ordinary three-dimensional space, for example, addition of two vectors or multiplication by a number always yields another *vector*. When the elements of some given set can be combined and multiplied by numbers in this way they are said to comprise a *linear vector space* or, when the elements are functions, a *function space*.

We assume a familiarity with the elementary properties of vectors in three dimensions and recall one or two of the main concepts. An arbitrary vector $\mathbf{r}$ can be expressed in terms of any three non-coplanar vectors $\mathbf{e}_1$, $\mathbf{e}_2$, $\mathbf{e}_3$ which define a *basis*:

$$\mathbf{r} = r_1\mathbf{e}_1 + r_2\mathbf{e}_2 + r_3\mathbf{e}_3 \tag{3.36}$$

---

‡For a fuller account of vectors, matrices and related matters see, for example, the books by Hall (1965) and Halmos (1942). Here we simply review the basic ideas and terminology.

§Note that equality of two functions means equality for all values taken by the argument.

In other words any *four* vectors v_1, v_2, v_3, v_4 (in this case r, e_1, e_2, e_3) must be connected by a linear relationship of the form

$$c_1 v_1 + c_2 v_2 + c_3 v_3 + c_4 v_4 = 0, \qquad (3.37)$$

where $c_1, \ldots, c_4$ are numerical coefficients. Any four vectors connected in this way (excluding the trivial case in which all coefficients are zero) are said to be *linearly dependent*; whereas any four vectors for which (3.37) *cannot* be satisfied, for any non-trivial choice of coefficients, are *linearly independent*. In three-dimensional space the *maximum* number of linearly independent vectors we can find is *three*, any fourth must then be linearly dependent on the first three, and may consequently be expressed in terms of them and in the form (3.36). The coefficients of e_1, e_2, e_3 in (3.36) are called the *components* of the vector r with respect to the basis $\{e_1, e_2, e_3\}$. The totality of all vectors of the form (3.36) constitutes a 3-*dimensional vector space* and the basis vectors are said to *span* the space. Three vectors are not *necessarily* linearly independent: if e_1, e_2, and e_3 happened to lie in the same plane a relationship would exist among them and we could then write e_3, say, in terms of e_1 and e_2. In general, e_1 and e_2 define a *two*-dimensional *sub*space: similarly, any vector e_1 spans a *one*-dimensional subspace—that comprising all vectors pointing along the line defined by e_1.

Similar ideas apply in a function space. Let us consider, for example, a set of three degenerate eigenfunctions, ϕ_1, ϕ_2, ϕ_3, of some operator H, with a common eigenvalue E. Clearly

$$\psi = c_1 \phi_1 + c_2 \phi_2 + c_3 \phi_3 \qquad (3.38)$$

is also an eigenfunction of H with eigenvalue E. By speaking of a three-fold degeneracy we mean that the maximum number of linearly independent eigenfunctions we can find is three and that any fourth must be expressible (in the above manner) in terms of the three already found. In other words no further *new* eigenfunctions can be found that do not reduce to linear combinations of ϕ_1, ϕ_2, ϕ_3. The functions ϕ_1, ϕ_2, ϕ_3 are *basis functions*, the coefficients c_1, c_2, c_3 are "components" of ψ with respect to the basis ϕ_1, ϕ_2, ϕ_3, and the totality of all functions of the form (3.38) constitutes a *three-dimensional function space*. There is clearly no restriction to spaces of three dimensions: a set of six degenerate eigenfunctions would span a six-dimensional function space provided any seventh eigenfunction with the same eigenvalue were found to be expressible as a linear combination of the six. In fact the existence of *complete sets* of functions (p. 48) suggests that the number of dimensions need not even remain finite. The ∞-dimensional linear space spanned by a complete set of quadratically integrable functions‡ is called a *Hilbert space*: the space spanned by the functions of a more restricted class (e.g. all eigenfunctions of an operator L with a given eigenvalue E) is then a *sub*space of Hilbert space. We shall find it very useful to develop further these geometrical analogies.

The scalar product

In ordinary three-dimensional space, the concepts of length and angle allow us to define the *scalar product* of two vectors

$$\mathbf{v} \cdot \mathbf{v}' = |\mathbf{v}|\,|\mathbf{v}'| \cos \theta = \mathbf{v}' \cdot \mathbf{v}, \qquad (3.39)$$

where $|\mathbf{v}|$, $|\mathbf{v}'|$ are the lengths of $\mathbf{v}$ and $\mathbf{v}'$ and θ is the angle between them. It is then convenient to introduce a basis $\mathbf{e}_1$, $\mathbf{e}_2$, $\mathbf{e}_3$ in which the vectors are mutually orthogonal and of unit length. Using the Kronecker delta, as in (3.17),

$$\mathbf{e}_i \cdot \mathbf{e}_j = \delta_{ij}. \qquad (3.40)$$

The scalar product $\mathbf{v} \cdot \mathbf{v}'$ is then easily expressed in terms of the *components* of the two vectors

$$\begin{aligned} \mathbf{v} \cdot \mathbf{v}' &= (v_1\mathbf{e}_1 + v_2\mathbf{e}_2 + v_3\mathbf{e}_3) \cdot (v_1'\mathbf{e}_1 + v_2'\mathbf{e}_2 + v_3'\mathbf{e}_3) \\ &= v_1v_1' + v_2v_2' + v_3v_3' \end{aligned}$$

where we have "multiplied out", noting that this is permitted because the geometrical definition shows that the product satisfies the usual associative and distributive laws. In the case $\mathbf{v}' = \mathbf{v}$ we obtain the squared length of $\mathbf{v}$:

$$|\mathbf{v}|^2 = \mathbf{v} \cdot \mathbf{v} = v_1{}^2 + v_2{}^2 + v_3{}^2.$$

It also follows that, with an orthonormal basis, any component of a vector may be expressed as a scalar product; by forming the scalar product of $\mathbf{v}$ with $\mathbf{e}_i$ and using (3.40) we obtain

$$v_i = \mathbf{e}_i \cdot \mathbf{v} \qquad (3.41)$$

and note that $v_i\mathbf{e}_i$ is the *projection* of the vector $\mathbf{v}$ in the direction of $\mathbf{e}_i$.

Again, similar ideas may be introduced in function space. In developing the theory in its general quantum mechanical context, the functions considered are usually orthonormal with *unit* weight factor (see p. 46): for real functions, which we consider first, the orthonormality property (3.17) suggests that we *define* a scalar product by

$$\phi_i \cdot \phi_j = \int_a^b \phi_i(x)\phi_j(x)\,dx = \delta_{ij}, \qquad (3.42)$$

where $\phi_i \cdot \phi_i = 1$ corresponds to "unit length" of a normalized func-

---

‡More precise definitions may be found in the books by Stone (1932) or von Neumann (1955).

tion, and $\phi_i \cdot \phi_j = 0$ $(i \neq j)$ to "perpendicularity" of two different orthogonal functions. From the definition of the scalar product as an integral, it follows easily that two n-term linear combinations,

$$\psi = \sum_{i=1}^{n} c_i \phi_i \quad \text{and} \quad \psi' = \sum_{j=1}^{n} c_j' \phi_j,$$

have a scalar product

$$\psi \cdot \psi' = c_1 c_1' + c_2 c_2' + \ldots + c_n c_n'. \tag{3.43}$$

The square of the "length" of a function is the *norm* $\|\psi\|^2$ defined as the sum of the squares of its components, referred to an orthonormal basis; and there is an exact parallel with the scalar product used in elementary vector algebra. Whether or not we can pass to the case $n \to \infty$ is essentially a question of whether or not such sums converge; the assumption that a set is complete, for quadratically integrable functions, ensures that they do.

A slight generalization is necessary when complex functions are admitted. In this case the important quantities are integrals such as (3.17) which contain the complex conjugate of one of the two functions involved. The norm of $\psi(x)$, again assuming unit weight factor in the integrand, is then

$$\|\psi\|^2 = \int_a^b \psi^*(x) \psi(x) \, dx$$

and is an essentially positive quantity only because one function is "starred". In view of these facts we define the *Hermitian scalar product* of functions $\psi(x)$, $\psi'(x)$ by

$$\psi^*\psi' = \int_a^b \psi^*(x)\psi'(x) \, dx = \langle \psi | \psi' \rangle, \tag{3.44}$$

where the symbolic notation on the left‡ stresses the formal analogy with elementary vector algebra, while that on the right is due to Dirac. The orthonormality property of basis functions is then

$$\langle \phi_i | \phi_j \rangle = \int_a^b \phi_i^*(x) \phi_j(x) \, dx = \delta_{ij} \tag{3.45}$$

and the main properties of the scalar product defined by (3.44) are, using α to denote an arbitrary complex number,

$$\begin{aligned} \langle \psi' | \psi \rangle &= \langle \psi | \psi' \rangle^*, \\ \langle \psi | \psi' + \psi'' \rangle &= \langle \psi | \psi' \rangle + \langle \psi | \psi'' \rangle, \\ \langle \psi | \alpha \psi' \rangle &= \alpha \langle \psi | \psi' \rangle, \\ \langle \psi | \psi \rangle &\geq 0 \end{aligned} \tag{3.46}$$

---

‡The dot may be omitted, the star between the function symbols (i.e. attached to the left-hand function) serving to indicate the scalar product. This notation (with a bar instead of the star) goes back to Condon and Shortley (1935). Other notations, such as (ψ, ψ'), are also commonly used. See also Morse and Feshbach (1953) Section 1.6.

The first equation expresses the "Hermitian symmetry" of the scalar product; interchanging the two functions is accompanied by complex conjugation, the order being immaterial only for *real* functions. The last expresses the fact that "length" is defined as a real *positive* quantity. The other equations show that scalar products of linear combinations can be expanded in the usual way (p. 56), though it should be noted that $\langle \alpha\psi|\psi'\rangle = \alpha^*\langle\psi|\psi'\rangle$. The properties embodied in (3.46) are characteristic of any Hilbert space, and are said to define its "metric".

Many of the metrical properties of ordinary three-dimensional space are shared by Hilbert space and many important theorems therefore express "common-sense" ideas. Thus, from the "metric axioms" (3.46), it may be shown that for any two functions ψ and ψ'

$$|\langle\psi|\psi'\rangle|^2 \leqq \langle\psi|\psi\rangle\langle\psi'|\psi'\rangle. \tag{3.47}$$

This is the famous "Schwarz inequality". It merely states that the scalar product of two vectors cannot exceed the product of their lengths, the equality being achieved when one is a multiple of the other. It follows from this result that the scalar product of any two quadratically integrable functions must always exist, i.e. the integral which defines it must converge.

Finally, we note that the properties of complete orthonormal sets are easily given a geometrical interpretation. Thus the formula (3.23) for the expansion coefficient c_i may be written

$$c_i = \langle\phi_i|\psi\rangle \tag{3.48}$$

which is merely the expression (3.6) for a vector component as a scalar product. The completeness property itself (3.24 et seq.) also takes a transparent form. If $\psi_n(x)$ is an n-term approximant to the function $\psi(x)$, the norm becomes, cf. (3.43),

$$||\psi_n||^2 = \langle\psi_n|\psi_n\rangle = |c_1|^2 + |c_2|^2 + \ldots |c_n|^2.$$

The fact that this cannot be greater than $\langle\psi|\psi\rangle$, so that convergence is from below, simply means that the vector ψ_n (in which components c_i with $i > n$ have been given zero values) is "shorter" than ψ—a result which is geometrically obvious in the example of Fig. 3.2. In general ψ_n is a *projection* of ψ on the n-dimensional subspace spanned by functions $\{\phi_1, \phi_2, \ldots \phi_n\}$ and the inequality (3.24) states that the projection of a vector on any subspace can never be longer than the vector itself.

All the preceding ideas may be extended without formal change from functions of one variable to functions of several variables, whenever suitable complete sets are available. The scalar product of two functions, Φ_i and Φ_j, will then usually be written

$$\langle\Phi_i|\Phi_j\rangle = \int\Phi_i^*(\mathbf{x})\Phi_j(\mathbf{x})\,d\mathbf{x} \tag{3.49}$$

where the bold letter **x** stands for all the variables concerned and $\int dx$ implies integration over the appropriate interval for each variable (e.g. over all space for the coordinates of an electron). In stating the axioms of quantum mechanics (Chapter 4) such sets are assumed to

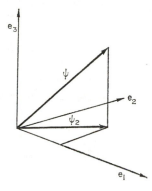

FIG. 3.2. Projection onto a subspace. The projection ψ_2, onto the two-dimensional subspace defined by e_1 and e_2, is obtained by removing the e_3 component of ψ. In general, ψ_2 is shorter than ψ (Bessel's inequality).

exist, and the representation of states by vectors in a Hilbert space plays a central role. The formulation also leans heavily on the theory of matrix representations to which we now turn.

3.5. Matrix representation of operators

Two vectors, **v** and **v′**, may differ in both magnitude and direction: **v′** may be produced from any given **v** by rotation and change of length‡ and it is convenient to write

$$\mathbf{v}' = \mathbf{R}\mathbf{v} \tag{3.50}$$

where the operator **R** indicates the operation performed on **v**. The vectors may be in ordinary space or may be elements of a function space; we often use the same terminology and note that, in either case, sums and products of operators may be defined as in (3.3) and (3.4). Thus, if rotation **B** followed by rotation **A** produces from **v** the same vector **v′** as a single rotation **C** (for *any* given **v**) we write **AB** = **C**. The *product* of two rotations is thus simply the single rotation which has exactly the same effect on any vector as the two rotations *performed*

---

‡In what follows we usually refer to such *generalized* rotations simply as rotations. All the operators we shall meet in quantum mechanics are linear (p. 42) and may be represented as rotations in a suitable vector space.

sequentially. Again, the order of the factors is important and it is not necessary that $BA = AB$. The *sum* of two rotations, A and B, is defined as the operation of applying A and B separately and forming the vector sum of the results; $(A+B)v = Av+Bv$. If the result is identical with Cv, for any choice of v, we write $(A+B) = C$. Here we shall be concerned mainly with the product and with the need to characterize an operator *numerically*, so that its effect on any given vector may be worked out by simple arithmetic.

Let us consider first the three-dimensional case, expressing the vectors v and v' in terms of three orthogonal unit vectors, e_1, e_2, e_3. Noting that rotation is a linear operator, so that rotation of the sum of two vectors gives the same result as summing the two rotated vectors, we may write (3.50) in the form

$$v_1'e_1+v_2'e_2+v_3'e_3 = R(v_1e_1+v_2e_2+v_3e_3)$$
$$= v_1(Re_1)+v_2(Re_2)+v_3(Re_3).$$

If we now take the scalar product with e_1 we obtain, since $e_i \cdot e_j = \delta_{ij}$,

$$v_1' = (e_1 \cdot Re_1)v_1+(e_1 \cdot Re_2)v_2+(e_1 \cdot Re_3)v_3$$

with similar expressions for v_2' and v_3'. The components of the rotated vector $v' = Rv$ are therefore expressible in terms of those of the original vector v; the coefficients involved are scalar products involving only the rotation operator and the unit vectors of the basis and therefore have the same values for *any* choice of v. The operator R is therefore completely characterized by its effect on the vectors of a basis—more specifically with the orientations (as determined by scalar products) of the rotated unit vectors relative to the original basis. If we put

$$R_{ij} = e_i \cdot Re_j \qquad (3.51)$$

the components of v' are related to those of v by

$$v_1' = R_{11}v_1+R_{12}v_2+R_{13}v_3,$$
$$v_2' = R_{21}v_1+R_{22}v_2+R_{23}v_3,$$
$$v_3' = R_{31}v_1+R_{32}v_2+R_{33}v_3.$$

We recall that such systems of equations may be written in a shorthand notation in two ways. In the *subscript* form we write

$$v_i' = \sum_{j=1}^{3} R_{ij}v_j \quad (i = 1, 2, 3) \qquad (3.52a)$$

(each value of i giving one of the foregoing equations), while in *matrix* form we group the coefficients R_{ij} into a square array and the components into columns, and write

$$\begin{bmatrix} v_1' \\ v_2' \\ v_3' \end{bmatrix} = \begin{bmatrix} R_{11} & R_{12} & R_{13} \\ R_{21} & R_{22} & R_{23} \\ R_{31} & R_{32} & R_{33} \end{bmatrix} \begin{bmatrix} v_1 \\ v_2 \\ v_3 \end{bmatrix} \tag{3.52b}$$

or more briefly,

$$\mathbf{v}' = \mathbf{R}\mathbf{v} \tag{3.52c}$$

where bold letters denote the arrays, or *matrices*, which are manipulated according to the rules of matrix algebra.‡

Equation (3.52c) is formally similar to (3.50) and is a *matrix representation* of that equation, the matrix $\mathbf{R}$ being associated with the operator R. The actual elements in the array R clearly depend on the choi e of a particular basis of unit vectors e_1, e_2, e_3, and this choice must therefore be indicated in defining any particular representation. We note also that the same array $\mathbf{R}$ appears in another context. If the rotation R is applied to the vectors of the basis it yields a rotated set of basis vectors e_1', e_2', e_3' and the components of e_1' $(= Re_1)$, *relative* to e_1, e_2, e_3, are $e_1 \cdot Re_1$, $e_2 \cdot Re_1$, $e_3 \cdot Re_1$: thus, from (3.51),

$$e_1' = e_1 R_{11} + e_2 R_{21} + e_3 R_{31} = \sum_k e_k R_{k1}$$

where the coefficients have been written on the right of the basis vectors simply to conform to the "chain rule" for matrix products (see footnote), matching indices coming next to each other. Similarly

$$e_2' = e_1 R_{12} + e_2 R_{22} + e_3 R_{32},$$

$$e_3' = e_1 R_{13} + e_2 R_{23} + e_3 R_{33}$$

and the whole set of equations determining rotation of the basis may then be collected into the subscript form

$$e_i' = Re_i = \sum_k e_k R_{ki} \quad (i = 1, 2, 3) \tag{3.53a}$$

---

‡Some familiarity with matrix notation is assumed. Here we need only note that an $m \times n$ ("m by n") matrix $\mathbf{R}$ is an array of m rows and n columns, R_{ij} being the "element" in the ith row and jth column; that equality of two matrices means equality of all corresponding pairs of elements; and that in a product $\mathbf{C} = \mathbf{AB}$ the element C_{ij} is obtained by taking the elements on the *ith row* of $\mathbf{A}$, multiplying by corresponding elements in the *jth column* of $\mathbf{B}$ and summing the products so formed; i.e. that

$$C_{ij} = \sum_k A_{ik} B_{kj}.$$

The product is only defined if the number of columns in $\mathbf{A}$ matches the number of rows in $\mathbf{B}$ and the matrices are then "conformable": if $\mathbf{A}$ is $p \times q$ and $\mathbf{B}$ is $q \times r$, the product is a $p \times r$ array. If there is only one row or column only one subscript is required and it must then be remembered that *components* (e.g. v_1, v_2, v_3) are conventionally collected into *columns* (i.e. matrices of several "rows", one column). In equation (3.52b) a 3×1 matrix is obtained as the product of a 3×3 times a 3×1.

or the matrix form

$$(e_1' \, e_2' \, e_3') = R(e_1 \, e_2 \, e_3) = (e_1 \, e_2 \, e_3) \begin{bmatrix} R_{11} \, R_{12} \, R_{13} \\ R_{21} \, R_{22} \, R_{23} \\ R_{31} \, R_{32} \, R_{33} \end{bmatrix} \quad (3.53b)$$

which may be written more briefly

$$\mathbf{e}' = R\mathbf{e} = \mathbf{e}R. \quad (3.53c)$$

Equations (3.53a, b, c), which refer to rotation of *basis vectors*, should be compared with (3.52a, b, c) which refer to *components* of an arbitrary rotated vector. In all such equations the position of the subscripts must be carefully observed; this requirement is taken care of automatically in the matrix forms, provided components of a vector are always written as *columns* and sets of basis vectors as *rows*. It should be noted that any vector may then be written as a row-column product of basis vectors and components:

$$\mathbf{v} = \sum_i v_i \, \mathbf{e}_i = \mathbf{e}\mathbf{v}. \quad (3.54)$$

If the above conventions are ignored or used improperly, considerable confusion may arise.

EXAMPLE. *Rotations in three dimensions.* Let us take e_3 as a rotation axis: if the basis is rotated through θ we then obtain

$$e_1' = \cos\theta\, e_1 + \sin\theta\, e_2, \quad e_2' = -\sin\theta\, e_1 + \cos\theta\, e_2, \quad e_3' = e_3$$

which corresponds to (3.53) with the association

$$R \rightarrow \mathbf{R} = \begin{bmatrix} \cos\theta & -\sin\theta & 0 \\ \sin\theta & \cos\theta & 0 \\ 0 & 0 & 1 \end{bmatrix},$$

i.e. with the rotation R we associate the matrix $\mathbf{R}$, which describes the rotation in terms of the basis **e**. For a general rotation we should obtain $e_i' = l_i e_1 + m_i e_2 + n_i e_3$ where l_i, m_i, n_i are the *direction cosines* of e_i' relative to e_1, e_2, e_3. The matrix describing the rotation is then

$$\mathbf{R} = \begin{bmatrix} l_1 & l_2 & l_3 \\ m_1 & m_2 & m_3 \\ n_1 & n_2 & n_3 \end{bmatrix}$$

each column containing the components of a rotated basis vector.

The central result of representation theory may now be stated as follows. If two operators, A and B, applied sequentially are equivalent to a third operator,

$$C = AB, \quad (3.55)$$

then the matrices $\mathbf{A}$, $\mathbf{B}$, $\mathbf{C}$ which represent the operators according to the definition (3.53), are related in an exactly similar way,

$$\mathbf{C} = \mathbf{A}\mathbf{B} \quad (3.56)$$

where **A** and **B** are combined by *matrix multiplication*. In mathematical language there is an *isomorphism* (identity of structure) between the set of operators, with sequential performance as the law of combination, and the set of matrices, with matrix multiplication as the law of combination. The proof of the equivalence is straightforward and may be found in any textbook on vectors and matrices.

EXAMPLE. *Composition of rotations.* Suppose A and B are rotations through θ_A and θ_B, respectively, about the e_3 axis. Multiplication of the matrices yields, with the abbreviation $s_A = \sin \theta_A$, $c_A = \cos \theta_A$, etc.,

$$\mathbf{AB} = \begin{bmatrix} c_Ac_B - s_As_B & -c_As_B - s_Ac_B & 0 \\ s_Ac_B + c_As_B & -s_As_B + c_Ac_B & 0 \\ 0 & 0 & 1 \end{bmatrix} = \begin{bmatrix} \cos(\theta_A + \theta_B) & -\sin(\theta_A + \theta_B) & 0 \\ \sin(\theta_A + \theta_B) & \cos(\theta_A + \theta_B) & 0 \\ 0 & 0 & 1 \end{bmatrix}$$

But this is evidently the matrix (**C**) associated with the rotation C = AB through the angle $(\theta_A + \theta_B)$. In other words, when the operations are combined by sequential performance their representative matrices must be combined by matrix multiplication. The present example is trivial but the result is general.

All the foregoing considerations may be applied without formal change to the case of operators in a Hilbert space, provided the scalar product is interpreted as the Hermitian scalar product (3.49), the "rotations" may then be, for example, the differential operators which produce a new function Ψ' from a given function Ψ; the basis is an orthonormal set of functions $\Phi_1, \Phi_2, \Phi_3, \ldots$ (in general an infinite set); and the matrices of the representation are infinite matrices. In place of the matrix element definition in (3.51) we then have, in full

$$R_{ij} = \langle \Phi_i | R | \Phi_j \rangle = \int \Phi_i^*(\mathbf{x}) R \Phi_j(\mathbf{x}) \, d\mathbf{x} \tag{3.57}$$

where we have written the scalar product $\langle \Phi_i | R\Phi_j \rangle$ with the more symmetrical notation usually adopted. The notation employed, in the quantum mechanical applications, is summarized in Table 3.1.

TABLE 3.1

Summary of matrix notation

Set of basis functions:	$\bar{\Phi} = (\Phi_1\Phi_2 \ldots \Phi_i \ldots)$	(row matrix)
Expansion of a function:	$\Psi = \Sigma_i \bar{\Phi}_i c_i = \bar{\Phi}\mathbf{c}$	(row-column product)
Rotation of basis:	$\bar{\Phi}' = R\Phi = \Phi R$	(R a square matrix representing R)
Rotation of arbitrary function:	$\Psi' = R\Psi$ $\mathbf{c}' = R\mathbf{c}$	(operator form) (matrix form)
Expansion coefficients in Ψ: Matrix elements of operator R:	$c_i = \langle \Phi_i | \Psi \rangle$ $R_{ij} = \langle \Phi_i | R | \Phi_j \rangle$	(valid only for orthonormal basis)

With the above conventions the matrices $\mathbf{A}$, $\mathbf{B}$, $\mathbf{C}$, ... associated with a set of operators in Hilbert space have exactly the same multiplicative properties as the operators themselves; the statements AB = C and $\mathbf{AB} = \mathbf{C}$ are exactly equivalent in the sense that each implies, and is implied by, the other. A representation of operators by matrices in this way is said to be *faithful* or *one-to-one* because it implies that different operators must have different matrices associated with them. In the quantum mechanical context it is normally assumed that there is a one-to-one correspondence between operators and the matrices (or other entities) which "represent" them. In mathematics, however (particularly in group theory, which we use in Vol. 2), the term "representation" is used in a wider sense: the matrices $\mathbf{A}$, $\mathbf{B}$, $\mathbf{C}$, ... associated with operators A, B, C, ... may have the same multiplicative properties (in the sense that AB = C requires $\mathbf{AB} = \mathbf{C}$) *without* the proviso that $\mathbf{A}$, $\mathbf{B}$, $\mathbf{C}$, ... shall all be distinct (e.g. the same matrix may be associated always with a *pair* of operators, in which case the representation is termed "two-to-one"). In this case AB = C implies $\mathbf{AB} = \mathbf{C}$ but the reverse is no longer true.

3.6. Change of representation

The particular matrix associated with any given operator depends on choice of the basis defining the representation. In Chapter 5 we shall need to know the relationship between the matrices arising from two different choices of basis.

Suppose we have two complete, linearly independent sets $\Phi = (\Phi_1 \Phi_2 \ldots \Phi_i \ldots)$ and $\overline{\Phi} = (\overline{\Phi}_1 \overline{\Phi}_2 \ldots \overline{\Phi}_i \ldots)$, which we assume are orthonormal and provide alternative bases for a representation. Each function $\overline{\Phi}_i$ may then be expressed in terms of the original Φ's (completeness property) and we write the relationship between the two bases as

$$\overline{\Phi}_i = \sum_{j=1}^{\infty} \Phi_j T_{ji} \quad (i = 1, 2, \ldots), \qquad (3.58a)$$

where the coefficients T_{ji} may be collected into a *transformation matrix* $\mathbf{T}$. In matrix notation the relationship becomes

$$\boxed{\overline{\Phi} = \Phi \, \mathbf{T}.} \qquad (3.58b)$$

The sets of expansion coefficients defining an arbitrary function with respect to the two alternative bases are then easily related; denoting the sets by $\mathbf{c}$ and $\bar{\mathbf{c}}$ (column matrices) the alternative expressions for Ψ are, using (3.58b),

$$\Psi = \Phi \mathbf{c} = \overline{\Phi}\bar{\mathbf{c}} = \Phi \mathbf{T}\bar{\mathbf{c}}$$

and equality of corresponding components then requires that the columns c and $T\bar{c}$ are element-by-element identical:

$$c = T\bar{c}. \tag{3.59}$$

Now T is a non-singular‡ matrix and thus possesses an *inverse*, denoted by T^{-1} with the property

$$TT^{-1} = T^{-1}T = 1 \tag{3.60}$$

where 1 is the unit matrix, consisting of 1's on the diagonal and 0's elsewhere, which leaves any other matrix unchanged in multiplication (i.e. $1M = M1 = M$). On multiplying both sides of (3.59) by T^{-1} we obtain the inverse relationship

$$\boxed{\bar{c} = T^{-1}c.} \tag{3.61}$$

Thus the expansion coefficients which represent the same function Ψ relative to two different bases follow a different transformation law (3.61) from the basis vectors themselves (3.58b). The fact that Ψ is *invariant* against change of description is nicely reflected in the identity of the alternative expressions: $\overline{\Phi}\bar{c} = \Phi TT^{-1}c = \Phi c$. Basis vectors and components transform "contragrediently".

We now consider the relationship between matrices A and $\bar{A}$ which describe the same operator A relative to different bases, Φ and $\overline{\Phi}$. The matrix A is defined by

$$A\Phi_i = \sum_j \Phi_j A_{ji}$$

or

$$A\Phi = \Phi A \tag{3.62a}$$

whereas the corresponding matrix describing A in the new basis appears in

$$A\overline{\Phi} = \overline{\Phi}\bar{A}. \tag{3.62b}$$

But since the bases are related by (3.58b), and hence by $\Phi = \overline{\Phi}T^{-1}$, substitution in (3.62a) yields

$$A\overline{\Phi}T^{-1} = \overline{\Phi}T^{-1}A$$

or, on multiplying by T,

$$A\overline{\Phi} = \overline{\Phi}(T^{-1}AT).$$

By comparison with (3.62) it is then clear that the matrix $\bar{A}$ representing A in the $\overline{\Phi}$ basis is

$$\bar{A} = T^{-1}AT. \tag{3.63}$$

‡The determinant of the matrix is non-zero. This is equivalent to saying that the bases Φ and $\overline{\Phi}$ are each linearly independent.

$\bar{\mathbf{A}}$ is said to be related to $\mathbf{A}$ by a *similarity transformation*. The reason for the term is clear, for if the operator relation $\mathsf{AB} = \mathsf{C}$ is equivalent to

$$\mathbf{AB} = \mathbf{C}$$

then multiplication from left and right by $\mathbf{T}^{-1}$ and $\mathbf{T}$, respectively, and insertion of the unit matrix $\mathbf{1} = \mathbf{TT}^{-1}$ between $\mathbf{A}$ and $\mathbf{B}$ on the left, yields

$$\bar{\mathbf{A}}\bar{\mathbf{B}} = \bar{\mathbf{C}}.$$

Any relationship among the matrices of one representation therefore implies an exactly similar relationship among those of any other.

Finally, a simplification occurs in the special case where only orthonormal bases are considered. The matrices relating such bases have a special property which follows on considering the transformation of the array of all basis vector scalar products; this array, which explicitly defines the "metric" of the space, occurs in the general scalar product expression‡

$$\langle \Psi | \Psi' \rangle = \sum_{i,j} c_i{}^* c_j{}' \langle \Phi_i | \Phi_j \rangle. \tag{3.64a}$$

This may be written in matrix form by introducing the *Hermitian transpose*§ $\mathbf{M}^\dagger$ of a matrix $\mathbf{M}$, obtained by interchanging rows and columns and taking the complex conjugate of every element: thus $M^\dagger{}_{ij} = M_{ji}{}^*$ and in particular $\mathbf{c}^\dagger$ is a *row* matrix with elements $c_i{}^*$. With this notation (3.64a) becomes

$$\langle \Psi | \Psi' \rangle = \mathbf{c}^\dagger \mathbf{M} \mathbf{c}', \tag{3.64b}$$

where $\mathbf{M}$ denotes the square matrix of scalar products (the "metric" or "overlap" matrix) with $M_{ij} = \langle \Phi_i | \Phi_j \rangle$.

To find how $\mathbf{M}$ and $\bar{\mathbf{M}}$ for two different bases are related, it is convenient to use the symbolic form $\langle \Phi_i | \Phi_j \rangle = \Phi_i{}^* \Phi_j$ and to represent each matrix as a column-row product: thus

$$\begin{bmatrix} \Phi_1{}^*\Phi_1 & \Phi_1{}^*\Phi_2 & \cdots \\ \Phi_2{}^*\Phi_1 & \Phi_2{}^*\Phi_2 & \cdots \\ \cdots & \cdots & \cdots \end{bmatrix} = \begin{bmatrix} \Phi_1{}^* \\ \Phi_2{}^* \\ \cdots \end{bmatrix} (\Phi_1 \Phi_2 \cdots)$$

or more briefly $\mathbf{M} = \mathbf{\Phi}^\dagger \mathbf{\Phi}$. For the second basis $\bar{\mathbf{\Phi}}$, the overlap matrix is

‡This follows from the axioms (3.46) applied to arbitrary linear combinations of the basis functions. The symbolic form is more transparent:
$$\Psi^* \Psi' = (c_1 \Phi_1 + c_2 \Phi_2 + \ldots)^*(c_1{}' \Phi_1 + c_2{}' \Phi_2 + \ldots) = \sum_{i,j} c_i{}^* c_j{}' \Phi_i{}^* \Phi_j.$$

§Also called Hermitian conjugate, adjoint, or associate matrix.

then given by

$$\overline{\mathbf{M}} = \overline{\boldsymbol{\Phi}}^{\dagger}\overline{\boldsymbol{\Phi}} = (\boldsymbol{\Phi}\mathbf{T})^{\dagger}(\boldsymbol{\Phi}\mathbf{T}) = \mathbf{T}^{\dagger}(\boldsymbol{\Phi}^{\dagger}\boldsymbol{\Phi})\mathbf{T} = \mathbf{T}^{\dagger}\mathbf{M}\mathbf{T}. \qquad (3.65)$$

Here we have used the fact that in taking the Hermitian transpose of a matrix product the matrices must be written in reverse order, as well as having the daggers added: $(\mathbf{AB})^{\dagger} = \mathbf{B}^{\dagger}\mathbf{A}^{\dagger}$. If now we require both bases to be orthonormal, so that $\mathbf{M}$ and $\overline{\mathbf{M}}$ are both unit matrices, (3.65) becomes $\mathbf{T}^{\dagger}\mathbf{T} = \mathbf{1}$. When $\mathbf{T}$ is a non-singular square matrix of finite dimension‡ this is equivalent to the statement $\mathbf{T}^{\dagger} = \mathbf{T}^{-1}$. It follows that the inverse matrix, encountered in (3.63), may be formed simply by Hermitian transposition, and that the transformation matrix connecting two unitary bases must satisfy

$$\mathbf{T}^{\dagger}\mathbf{T} = \mathbf{T}\mathbf{T}^{\dagger} = \mathbf{1} \qquad (3.66)$$

A matrix with this property is said to be *unitary*, and the corresponding form of (3.63), namely

$$\bar{\mathbf{A}} = \mathbf{T}^{\dagger}\mathbf{A}\mathbf{T} \qquad (3.67)$$

defines a *unitary transformation*. The rotation *operator* T with which T is associated has a property analogous to (3.66). $\mathsf{T}^{\dagger}\mathsf{T} = 1$, and is also said to be unitary.

We summarize below the equations connecting different bases:

	$\boldsymbol{\Phi}$ *representation*	$\overline{\boldsymbol{\Phi}}$ *representation* $(\overline{\boldsymbol{\Phi}} = \boldsymbol{\Phi}\mathbf{T})$
$\Psi \rightarrow$	$\mathbf{c}$	$\bar{\mathbf{c}} = \mathbf{T}^{-1}\mathbf{c}$
$\mathsf{A} \rightarrow$	$\mathbf{A}$	$\bar{\mathbf{A}} = \mathbf{T}^{-1}\mathbf{A}\mathbf{T}$

(3.68)

If the bases are orthonormal, $\overline{\boldsymbol{\Phi}} = \boldsymbol{\Phi}\mathbf{U}$, these correspondences become

$\Psi \rightarrow$	$\mathbf{c}$	$\bar{\mathbf{c}} = \mathbf{U}^{\dagger}\mathbf{c}$
$\mathsf{A} \rightarrow$	$\mathbf{A}$	$\bar{\mathbf{A}} = \mathbf{U}^{\dagger}\mathbf{A}\mathbf{U}$

(3.69)

Such transformations are used extensively in Chapter 5.

3.7. Hermitian operators and eigenvalue equations in vector space

Since the eigenvalue equation plays such an important role in quantum mechanics it is useful to make a special study of the form which it takes on setting up a matrix representation. We introduce any orthonormal basis $\{\boldsymbol{\Phi}\}$ and at once transcribe the operator equation

$$\mathsf{H}\Psi = E\Psi \qquad (3.70)$$

into the *matrix* form‡

$$\mathbf{Hc} = E\mathbf{c}. \qquad (3.71a)$$

---

‡It has been tacitly assumed so far that there is no difficulty in passing to the limit where all matrices are infinite. We continue to ignore questions of mathematical rigor.

Here $\mathbf{H}$ is an infinite matrix with elements $H_{ij} = \langle \Phi_i | \mathsf{H} | \Phi_j \rangle$ and $\mathbf{c}$ is an infinite column of expansion coefficients, expressing the function Ψ in the form

$$\Psi = \sum_i c_i \Phi_i.$$

The matrix $\mathbf{H}$ has a special form because it represents a Hermitian operator. In general, the matrix elements of an operator A and its adjoint $\mathsf{A}^\dagger$ are related (according to (3.31)) by

$$\langle \Phi_i | \mathsf{A}^\dagger | \Phi_j \rangle = \langle \mathsf{A}\Phi_i | \Phi_j \rangle = \langle \Phi_j | \mathsf{A} | \Phi_i \rangle^*$$

The matrix representing $\mathsf{A}^\dagger$ is thus obtained from that representing A by interchanging rows and colums and taking the complex conjugate of every element; but this yields the Hermitian transpose, already indicated by adding a dagger to the matrix, and the notation is thus consistent—if A is represented by $\mathbf{A}$, the adjoint operator $\mathsf{A}^\dagger$ is represented by the Hermitian transpose $\mathbf{A}^\dagger$. If an operator is Hermitian (and thus *self*-adjoint) its matrix will also have the Hermitian symmetry property $\mathbf{A}^\dagger = \mathbf{A}$. The matrix in (3.71a) is thus Hermitian symmetric, $H_{ji} = H_{ij}^*$, or simply a *Hermitian matrix*.

To avoid a direct discussion of infinite matrices we assume the basis is truncated after n functions, where n is large but finite. Later (Vol. 2) we shall find that results obtained using a truncated expansion (the normal practice in most approximation methods) converge upon the exact results for $n \to \infty$. In the finite case (3.71a) takes the explicit form

$$\begin{bmatrix} H_{11} & H_{12} & \dots & H_{1n} \\ H_{21} & H_{22} & \dots & H_{2n} \\ \cdot & \cdot & \cdots & \cdot \\ H_{n1} & H_{n2} & \dots & H_{nn} \end{bmatrix} \begin{bmatrix} c_1 \\ c_2 \\ \cdot\cdot \\ c_n \end{bmatrix} = E \begin{bmatrix} c_1 \\ c_2 \\ \cdot\cdot \\ c_n \end{bmatrix} \tag{3.71b}$$

The corresponding simultaneous equations (namely, $H_{11}C_1 + H_{12}C_2 + \dots + H_{1n}C_n = Ec_1$, etc.) are the *secular equations*. For an arbitrary value of E they cannot all be satisfied because, on dividing throughout by, say, c_1, we have n equations to determine only $(n-1)$ unknown ratios (c_2/c_1), (c_3/c_1), $\dots$ (c_n/c_1): the ratios could be determined from any $(n-1)$ of the equations and the nth would not necessarily be satisfied. In reality, however, E is the nth unknown: by suitably choosing the value of E we can assume that the nth equation *is* satisfied by the ratios determined from the other $(n-1)$ equations. In this case the equations are said to be *compatible*. It is shown in textbooks on algebra (e.g. Archbold (1961); see also Margenau and Murphy, 1943) that the condition for compatibility is

$$\Delta(E) = \begin{vmatrix} (H_{11}-E) & H_{12} & \dots & H_{1n} \\ H_{21} & (H_{22}-E) & \dots & H_{2n} \\ \dots & \dots & \dots & \dots \\ H_{n1} & H_{n2} & \dots & (H_{nn}-E) \end{vmatrix} = 0 \qquad (3.72a)$$

or, in abbreviated form,

$$\Delta(E) = \det \left| H_{ij} - E\delta_{ij} \right| = 0. \qquad (3.72b)$$

The *secular determinant* $\Delta(E)$ expands to give an nth degree polynomial in E; this has n roots, which are the n eigenvalues of the matrix $\mathbf{H}$, and if any of these is substituted back into (3.71b) we can solve $(n-1)$ equations for the *ratios* of the coefficients, and then fix their absolute values by normalization. Thus for each eigenvalue E_K we can obtain an eigenvector‡ $\mathbf{c}_K$. As the set $\{\Phi_i\}$ approaches completeness $(n \to \infty)$ the roots E_K and columns of expansion coefficients $\mathbf{c}_K$ should yield exact energies and wave functions of the Schrödinger equation (3.70). On the other hand, by introducing a basis, the original *operator* equation has been replaced by a *matrix* eigenvalue equation, which for many purposes is advantageous.

We now state some of the more important properties of matrix eigenvalue equations.

(i) Eigenvectors belonging to different eigenvalues are orthogonal in the sense

$$\mathbf{c}_K^\dagger \mathbf{c}_L = 0 \quad (E_K \neq E_L).$$

In case of degeneracy $(E_K = E_L)$ it is possible to form new combinations of $\mathbf{c}_K$ and $\mathbf{c}_L$ which *are* orthogonal. It is also possible to normalize any eigenvector and we may therefore assume without loss of generality

$$\mathbf{c}_K^\dagger \mathbf{c}_L = \delta_{KL}. \qquad (3.73)$$

The proof extends easily to multiple degeneracies (cf. p. 47).

(ii) Since any eigenvector satisfies (3.71a) it follows, multiplying from the left by $\mathbf{c}_K^\dagger$ and using the orthonormality property, that

$$\mathbf{c}_K^\dagger \mathbf{H} \mathbf{c}_K = E_K, \quad \mathbf{c}_K^\dagger \mathbf{H} \mathbf{c}_L = 0. \qquad (37.4)$$

These results express the properties (verified by using $\Psi_K = \sum_i c_{Ki}\Phi_i$, etc.)

‡For brevity it is usual to refer to the column of expansion coefficients, which determine Ψ_K, as an eigenvector of the matrix $\mathbf{H}$.

$$\langle \Psi_K | H | \Psi_K \rangle = E_K, \quad \langle \Psi_K | H | \Psi_L \rangle = 0 \qquad (3.75)$$

where, in the limit $n \to \infty$, E_K coincides with an eigenvalue of the operator equation (3.70). Thus, on using the eigenfunctions as a basis, H is represented by a *diagonal* matrix of eigenvalues.

(iii) The results in (i) and (ii) may be neatly expressed in terms of a square matrix, whose columns are the eigenvectors $c_1, c_2, \ldots c_K, \ldots$. On putting

$$C = (c_1 | c_2 | \ldots) \qquad (3.76)$$

and using

$$E = \text{diag}(E_1, E_2, \ldots)$$

to denote the diagonal matrix of eigenvalues, we obtain‡

$$E = C^\dagger HC, \quad C^\dagger C = 1. \qquad (3.77)$$

Thus C is a unitary matrix (p. 67) and according to (3.77) defines a transformation from the basis $\{\Phi_i\}$ to a new basis—that provided by the eigenfunctions $\{\Psi_K\}$—in which H is replaced by a matrix of diagonal form. Solution of an eigenvalue problem therefore amounts to reducing a matrix to diagonal form by a suitable unitary transformation.

The eigenvalues and vectors appearing above only coincide with the exact solutions of the *operator* equation (3.70) in the limit $n \to \infty$; but all the results listed above are valid for the matrix eigenvalue equations (3.71b) with n finite. The function sets $\{\Phi_i\}$ and $\{\Psi_K\}$ may be regarded simply as alternative bases in the same n-dimensional vector space.

Finally, we note that extension to *non*-orthogonal bases is sometimes necessary. The eigenvalue equation (3.71a) is then replaced by

$$Hc = EMc \qquad (3.78)$$

while the consistency condition (3.72b) becomes

$$\Delta(E) = \det |H_{ij} - E M_{ij}| = 0. \qquad (3.79)$$

The results (i)–(iii) are modified accordingly. Orthogonality is now with respect to a metric M, as in (3.64b), and instead of (3.73) we have

$$c_K^\dagger M c_L = \delta_{KL}$$

and equations (3.77) are replaced by

$$E = C^\dagger HC, \quad C^\dagger MC = 1. \qquad (3.80)$$

---

‡Note that partitioned matrices such as (3.76) may be multiplied as if the individual blocks—provided they are properly conformable (see p. 61)—were single numbers. Thus $C^\dagger C$ resembles a column-row product in which the resultant elements are $c_K^\dagger c_L$. Similarly $E = C^\dagger HC$ has elements $c_K^\dagger H c_L$.

The other results are unaffected. This brief discussion of eigenvalue equations is sufficient for most of what follows. The eigenvectors of an operator have, however, a very profound significance. In particular they allow us to introduce the "normal form" of an operator: for completeness this development is indicated in Appendix 4.

3.8. Composition of vector spaces. Product space

In Section 3.3 we remarked that from two complete orthonormal sets of functions $\{u_i(x)\}$ and $\{v_j(y)\}$ it was possible to construct a third set, complete for functions of both variables, x and y. Using the language of vector space theory, the totality of all product functions

$$w_{ij}(x, y) = u_i(x)v_j(y) \tag{3.81}$$

is said to span a *product space*. Such spaces may be defined formally for any two sets of vectors: if $\{u_i\}$ is m-dimensional, and $\{v_j\}$ is n-dimensional, then there will be mn products

$$u_1v_1, \ u_1v_2, \ \ldots \ u_2v_1, \ u_2v_2, \ \ldots \ u_mv_n$$

which together define an mn-dimensional vector space.

The most general operator R in product space will send any particular product into a new linear combination of products (cf. (3.53a))

$$R(u_kv_l) = \sum_{i,j} u_iv_jR_{ij,kl} \tag{3.82}$$

where $R_{ij,kl}$ is an element of an $mn \times mn$ matrix **R**, each row and column labelled by a double index. The matrix elements can be written as scalar products (cf. (3.57))

$$R_{ij,kl} = \langle u_iv_j|R|u_kv_l\rangle \tag{3.83}$$

and the results of previous sections may be taken over with no essential change.

In the quantum mechanical context, $\{u_i\}$ and $\{v_j\}$ are frequently complete function sets, $\{\psi_i(\mathbf{x}_1)\}$ and $\{\psi_j(\mathbf{x}_2)\}$ each involving the variables (symbolized by $\mathbf{x}_i$) of a different particle, and the completeness of the products means simply that an exact two-particle wave function can be expanded in the form

$$\Psi(\mathbf{x}_1, \mathbf{x}_2) = \sum_{ij} c_{ij}\psi_i(\mathbf{x}_1)\psi_j(\mathbf{x}_2). \tag{3.84}$$

We shall have occasion to use the product space concept in later chapters (more particularly in Vol. 3 where we deal with the construction of many-electron wave functions) but no further mathematical developments are necessary at this point, and we are now ready to set up the general formalism of quantum mechanics.

REFERENCES

ARCHBOLD, J. W. (1961) *Algebra*, Pitman, London.

CONDON, E. U. and SHORTLEY, G. H. (1935) *The Theory of Atomic Spectra*, Clarendon Press, Oxford.

COURANT, R. and HILBERT, D. (1953) *Methods of Mathematical Physics*, Vol. I, Interscience, New York.

DIRAC, P. A. M. (1947) *The Principles of Quantum Mechanics*, 3rd ed., Clarendon Press, Oxford.

HALL, G. G. (1965) *Vectors, Matrices and Tensors*, Pergamon Press, Oxford.

HALMOS, P. R. (1942) *Finite Dimensional Vector Spaces*, Princeton University Press, Princeton. (See also, for a useful short review, Powell, J. L. and Crasemann (1961) *Quantum Mechanics*, Addison Wesley, Reading, Mass.)

KEMBLE, E. C. (1937) *The Fundamental Principles of Quantum Mechanics*, McGraw Hill, New York (reprinted 1958 by Dover, New York).

MARGENAU, H. and MURPHY, G. M. (1943) *The Mathematics of Physics and Chemistry*, Van Nostrand, Toronto, New York and London.

MORSE, P. M. and FESHBACH, H. (1953) *Methods of Theoretical Physics* (2 Vols), McGraw Hill, New York, Toronto and London.

STONE, M. H. (1932) *Linear Transformations in Hilbert Space*, American Mathematical Society, Vol. 15 of Colloquium Publications, New York.

VON NEUMANN, J. (1955) *Mathematical Foundations of Quantum Mechanics*, Princeton University Press, Princeton (translated from the German editon by R. T. Beyer).

CHAPTER 4

GENERAL FORMULATION OF QUANTUM MECHANICS

4.1. The postulates

The whole theoretical framework of quantum mechanics may be built upon a small number of postulates, which absorb the ideas introduced in Chapters 1 and 2 and provide a better basis for further generalization. There is no unique set of postulates, and any set may be stated in any number of different "languages": we shall try to make the simplest possible statement that will meet our needs, freely using the Schrödinger language which is by now familiar and disregarding questions of mathematical rigour. We use the term "language", following Tolman (1938), to indicate a particular explicit realization of the quantum mechanical operators (e.g. as differential operators working on wave functions). In the general formulation the operators are indicated by symbols, exhibiting certain laws of combination and certain relationships such as "commutation rules", but their mathematical nature needs no further specification; in this sense the content of the operator statements does not depend on any particular, more concrete, realization that might be chosen. The more commonly used word "representation" has more specialized connections and will here be used less liberally.

Each postulate will take the form of a very general mathematical statement, whose content will then be explained in Schrödinger language. From the postulates certain consequences—or corollaries—follow; but the postulates and the corollaries will be discussed as they arise and illustrated in terms of familiar examples. The possibility of translation into other languages (transformation theory) will be taken up in a following chapter, along with the more detailed statistical interpretation of quantum mechanics. Many of the classic textbooks on the subject may usefully be consulted for further discussion of basic principles; we mention only those by Kemble (1937), Dirac (1947), von Neumann (1955), Schiff (1949) and the more recent, meticulous and comprehensive work by Messiah (1961).

4.2. The state vector and its time development

The first postulate formalizes the concept of *state* and provides the basic "equation of motion":

POSTULATE 1. The state of a system is completely determined by a "state vector" Ψ which develops in time according to

$$\mathsf{H}\Psi = i\hbar(\partial\Psi/\partial t) \qquad (4.1)$$

where H is the *Hamiltonian operator* associated with the classical Hamiltonian function H. This equation applies *except* at the instant of observation, intervention of an observer being assumed to produce discontinuous and unpredictable changes.

The terms "vector" and "operator" are here used in the widest sense. The vector is an element of a Hilbert space and the operator is a prescription leading from one element to another—notions dealt with in Chapter 3 in sufficient detail for present purposes. In the Schrödinger language, Ψ is a function of time and of the variables ($\mathbf{x}$) describing the positions of the particles comprising the system and the postulate is simply a general statement of Schrödinger's equation, written in (1.19) for a one-particle system. We continue to use capital letters for state vectors (or wave functions) appearing in *general* statements, applying to one or *many*-particle systems. Lower-case letters will apply specifically to *one*-particle wave functions.

The classical Hamiltonian H is in general the function that expresses the energy of the system in terms of position and momentum variables, and the operator H is obtained using the prescription referred to in (1.22). The form of H, in Schrödinger language, has been given already for a one-particle system in (1.21).

Two corollaries follow at once. The first may be stated in the form

---

COROLLARY 1. If Ψ satisfies the basic equation (4.1), then so does $c\Psi$ where c is any (constant) complex number.

---

and is simply a consequence of the fact that H, along with the other operators of quantum mechanics, has the property of *linearity* (stated in (3.2)). The vector describing the state of a system is therefore arbitrary to within a multiplicative factor (in general any complex number); the magnitude of this factor (but not its phase) may conventionally be fixed so that the state vector is a *unit* vector. In the Schrödinger formulation, any two wave functions Ψ_1 and Ψ_2, have a *Hermitian scalar product* defined, as in (4.44), as

$$\langle \Psi_1 | \Psi_2 \rangle = \int \Psi_1^*(\mathbf{x};t)\Psi_2(\mathbf{x};t)d\mathbf{x}$$

and the "unit vector" condition corresponds to normalization of the wave function (for all values of the time t):

$$\langle \Psi | \Psi \rangle = \int \Psi^*(\mathbf{x};t)\Psi(\mathbf{x};t)d\mathbf{x} = 1.$$

The meaning of this normalization (anticipated in Section 1.3) is discussed after stating Postulate 2.

The second corollary refers to a certain type of system—the analogue of a "conservative system" in classical dynamics:

COROLLARY 2. When H does not depend on time, (4.1) has special solutions representing *stationary states*

$$\Psi = \Phi e^{-iEt/\hbar} \qquad (4.2)$$

in which the time appears only in the "phase factor" $e^{-iEt/\hbar}$, provided the "amplitude factor" Φ satisfies the time-independent equation

$$H\Phi = E\Phi. \qquad (4.3)$$

This corollary was established and discussed in the Schrödinger formulation in Chapter 1, where Ψ was a function of t and the variables $\mathbf{x}$, and H was a differential operator; the general proof is formally the same. At present E is merely a "separation constant" with the dimensions of energy: its meaning follows from the next postulate.

4.3. The expectation value postulate

In Section 1.2 it was observed that the wave function for a stationary state, with a definite quantized value of the energy E, satisfied an eigenvalue equation of the form (4.3) with the energy as eigenvalue. In such a state the energy is an *observable* with a strictly reproducible value. But what can be said of other observables (e.g. dynamical variables such as momentum components) whose measured values may exhibit a "scatter" or "uncertainty"?

The next postulate provides an answer to this question by giving a rule for predicting the arithmetic mean value of any observable A, for a system in a given state Ψ, which would be obtained from a large number of observations conducted under identical conditions. As this average value is used in a predictive sense, it is usually referred to as the "expectation value" of A in state Ψ. In general the observed values will exhibit a scatter about the average, the existence of this "uncertainty" being a central feature of quantum theory: on the other hand, the existence of states in which one or more observables (e.g. the energy) have *definite* values (the uncertainty vanishing) clearly must not be precluded. This possibility is discussed presently.

The postulate itself is remarkably simple:

POSTULATE 2. With every physical observable A may be associated a Hermitian operator A such that‡

$$\langle A \rangle = \langle \Psi | A | \Psi \rangle \qquad (4.4)$$

where $\langle A \rangle$ denotes the expectation value of A at time t in the state described by the normalized state vector Ψ.

We note that Hermitian symmetry of an operator (p. 45) means $\langle \Psi_1 | A | \Psi_2 \rangle = \langle A\Psi_1 | \Psi_2 \rangle$, where the first form is the scalar product of Ψ_1 with $A\Psi_2$, the second being that of $A\Psi_1$ with Ψ_2. The basic property $\langle \Psi | \Psi' \rangle = \langle \Psi' | \Psi \rangle^*$ then allows us to infer that all expectation values of physical observables are real (cf. p. 52).

In the Schrödinger language, (4.4) becomes

$$\langle A \rangle = \langle \Psi | A | \Psi \rangle = \int \Psi^*(\mathbf{x};t)A\Psi(\mathbf{x};t)\,d\mathbf{x} \qquad (4.5)$$

the integral being the Hermitian scalar product of Ψ with $A\Psi$. For a one-particle system, the operator A may be set up by taking the classical expression for A in terms of Cartesian position and momentum variables (x, y, z, p_x, p_y, p_z) and then replacing each momentum component by a differential operator in accordance with the scheme suggested in Section 1.2, namely

$$p_x \to \mathsf{P}_x = \frac{\hbar}{i}\frac{\partial}{\partial x}, \quad p_y \to \mathsf{P}_y = \frac{\hbar}{i}\frac{\partial}{\partial y}, \quad p_z \to \mathsf{P}_z = \frac{\hbar}{i}\frac{\partial}{\partial y}. \qquad (4.6)$$

It is easily verified that these operators are Hermitian. The position variables (x, y, z) are left unchanged in making the association (4.6) and act merely as multiplying factors in the resultant operator. The generalization to a many-particle system is straightforward, each variable then carrying a subscript to distinguish the particles $1, 2, \ldots N$.

---

‡Recall that the second vertical stroke in (4.4) is inserted only to formally separate the three factors involved $\langle \text{vector} | \text{operator} | \text{vector} \rangle$. Thus $\langle A \rangle$ may be written equally well as $\langle A \rangle = \langle \Psi | A\Psi \rangle$, a scalar product of the vectors Ψ and $A\Psi$.

It must be noted that ambiguities may arise because, for example, xp_x and $p_x x$ are not distinguished in classical mechanics, whereas the associated operators $(\hbar/i)x(\partial/\partial x)$ and $(\hbar/i)(\partial/\partial x)x$ are quite different in their effects. These ambiguities may usually be eliminated by noting the requirement of Hermitian symmetry and using a symmetrized combination of the alternative expressions.

Before considering particular examples to illustrate the implications of Postulate 2, we establish two further corollaries. The first concerns the interpretation of the separation parameter E in (4.2) and (4.3). According to Postulate 2, the energy at any instant has an expectation value (since H is used to denote the operator associated with E)

$$\langle E \rangle = \langle \Psi | H | \Psi \rangle.$$

On multiplying each side of (4.3) by the phase factor in (4.2), and taking a scaler product with Ψ (from the left), the expectation value may evidently be written

$$\langle E \rangle = \langle \Psi | H | \Psi \rangle = E \langle \Psi | \Psi \rangle = E.$$

The implication is thus:

COROLLARY 3. The parameter E appearing in the stationary state equations (4.2) and (4.3) is the expectation value of the energy of the system.

We shall verify later that in a stationary state the energy is *definite* or *sharp* in the sense that any number of measurements, on a system in that given state, would yield the same result, i.e. no deviations would be found.

The second result specifically refers to the Schrödinger wave function but is of such fundamental importance that we include it among the corollaries. We consider, first for one particle only, the expectation value of any function of position, $f(x, y, z)$. This is, in a state with wave function ψ,

$$\langle f \rangle = \int \psi^*(x, y, z; t) f(x, y, z) \psi(x, y, z; t) \, dx \, dy \, dz$$

or, since $f(x, y, z)$ is merely a multiplying factor in the integrand,

$$\langle f \rangle = \int f(x, y, z) |\psi(x, y, z; t)|^2 dx \, dy \, dz. \tag{4.7}$$

The form of this expression justifies the interpretation

$$|\psi(x, y, z; t)|^2 dx \, dy \, dz = \begin{bmatrix} \text{probability of particle} \\ \text{being found in volume} \\ \text{element } dx \, dy \, dz \text{ at time } t \end{bmatrix}$$

—for the classical average value of any function of the position would be obtained by weighting $f(x, y, z)$ with the relative frequency of occurrence (i.e. probability) of the values x, y, z and "summing" (i.e. in this case integrating) over all possible points x, y, z, exactly as in (4.7). This is the statistical interpretation of the wave function first proposed by Born and already introduced briefly in Chapter 1.

The generalization to a many-particle system is straightforward: for brevity we let $\mathbf{x}_i$ stand for the variables of particle i, and state the result in the form

---

COROLLARY 4. For a many-particle system, the Schrödinger wave function has the significance

$$\left|\Psi(\mathbf{x}_1, \mathbf{x}_2, \ldots, \mathbf{x}_N; t)\right|^2 d\mathbf{x}_1 d\mathbf{x}_2 \ldots d\mathbf{x}_N = \begin{bmatrix} \text{probability of particle 1} \\ \text{in } d\mathbf{x}_1, \text{ particle 2} \\ \text{simultaneously in} \\ d\mathbf{x}_2, \text{ etc., at time } t \end{bmatrix}$$

(4.8)

where $\mathbf{x}_i$ denotes collectively all the (Schrödinger) variables characterizing particle i.

The expectation value of any function of the particle coordinates is given by an expression exactly analogous to (4.7).

To illustrate the physical content of Postulate 2 we now consider two simple but instructive examples of expectation value calculations.

EXAMPLE. *Linear momentum*. We use the Schrödinger wave functions obtained in Sections 2.1, 2.4 and calculate the expectation value of a linear momentum component p_x.

First we consider the free particle (Section 2.4) with wave function normalized within a cube of volume L^3; with the usual abbreviation‡ $\mathbf{r} = (x, y, z)$

$$\psi_k(\mathbf{r}; t) = \phi_k(\mathbf{r}) \exp(-iEt/\hbar)$$

$$\phi_k(\mathbf{r}) = L^{-3/2} \exp i(k_x x + k_y y + k_z z).$$

It is clear that the time-dependent phase factor may be ignored, cancelling with its complex conjugate in the integrand of (4.5) and that y- and z-dependent factors disappear in a similar way. The result is

$$\langle p_x \rangle = L^{-3} \int \exp(-ik_x x) \frac{\hbar}{i} \frac{\partial}{\partial x} \exp(ik_x x) \, dx \, dy \, dz$$

$$= L^{-1} \int \exp(-ik_x x) \frac{\hbar}{i} ik_x \exp(ik_x x) \, dx$$

$$= \hbar k_x.$$

---

‡Note that $\mathbf{x}$ has been used to indicate *all* variables, while $\mathbf{r}$ indicates only *spatial* variables. Spin is introduced in Section 4.9.

This is consistent with our interpretation of the current flow expression (p. 37) which suggested that k_x measured the x-component of linear momentum, in units of $\hbar$. The present result shows that $\hbar k_x$ is the *average* value which would be obtained in a large number of observations. The possibility of the momentum being definite (no deviations) is discussed presently.

Next let us consider the particle confined within a cube of side L, with wave function $\psi(\mathbf{r};t) = \phi(\mathbf{r})e^{-iEt/\hbar}$ where $\phi(\mathbf{r})$ is given by (2.9a). Again the time factor may be discarded, and the average momentum is

$$\langle p_x \rangle = (2/L)^{3/2} \int \left(\sin^2 \frac{\pi n_z}{L} \sin^2 \frac{\pi n_y}{L} \right) \sin \frac{\pi n_x}{L} \frac{\hbar}{i} \frac{\partial}{\partial x} \sin \frac{\pi n_x}{L} dx\,dy\,dz$$

which easily yields

$$\langle p_x \rangle = 0.$$

In other words, although the energy expression (2.9b) is still consistent with a momentum component of *magnitude* $\hbar k_x (k_x = \pi n_x/L)$, the *average* momentum vanishes, indicating equal likelihood of positive and negative momenta. The new boundary conditions correspond to a finite box with perfectly reflecting walls, between which the particle moves back and forth.

EXAMPLE. *Angular momentum.* In classical mechanics the x component of the angular momentum of a particle, about a given origin is defined by

$$L_x = yp_z - zp_y.$$

The corresponding Schrödinger operator is ‡

$$\frac{\hbar}{i} \left(y \frac{\partial}{\partial z} - z \frac{\partial}{\partial y} \right).$$

In the theory of angular momentum it is particularly convenient to use dimensionless quantities, and we therefore associate with L_x the operator

$$\mathsf{L}_x = \frac{1}{i} \left(y \frac{\partial}{\partial z} - z \frac{\partial}{\partial y} \right).$$

We must then remember that $\hbar \mathsf{L}_x$ corresponds to the actual angular momentum, and that the expectation value of L_x will give the angular momentum *in units of $\hbar$*, this being the corresponding *atomic unit* (p. 33).

Let us consider first any spherically symmetrical wave function, such as that describing the ground state of the electron in the hydrogen atom. The time factor may again be discarded, and we obtain

$$\langle \psi | \mathsf{L}_x | \psi \rangle = \int \phi^*(\mathbf{r}) \frac{1}{i} \left(y \frac{\partial}{\partial z} - z \frac{\partial}{\partial y} \right) \phi(\mathbf{r}) d\mathbf{r}.$$

Now in terms of the polar coordinates,§ r, θ, ϕ,

$$\frac{\partial}{\partial z} = \frac{\partial r}{\partial z} \frac{\partial}{\partial r} + \frac{\partial \theta}{\partial z} \frac{\partial}{\partial \theta} + \frac{\partial \phi}{\partial z} \frac{\partial}{\partial \phi}$$

---

‡There is clearly no ambiguity (cf. p. 77) since the operators commute, $(\partial/\partial z)(y\psi) = y(\partial/\partial z)\psi$.

§Use of ϕ for both an angle and the wave function is regretable, but the meaning is always clear from the context.

and for a spherically symmetrical function the last two terms vanish while the first becomes $(z/r)\,\partial/\partial r$. Similarly $\partial/\partial y$ is equivalent to $(y/r)\,\partial/\partial r$. The integrand therefore vanishes identically and $\langle\psi|\mathsf{L}_x|\psi\rangle = 0$.

It is not difficult to show that the expectation value of any angular momentum component vanishes for all *real* wave functions as a result of the Hermitian character of the operator. On the other hand, it may easily be verified that

$$(x+iy)f(r)$$

describes a state with one unit of angular momentum about the z-axis. Wave functions of this kind appear among the general solutions of (2.20) and are discussed in Part II.

4.4. Significance of the eigenvalue equation

We now turn to the most important corollary following from the expectation value postulate. This concerns the possibility of finding a state Ψ in which some quantity, A say, may have a perfectly definite or sharp value (p. 77) characteristic of that state and measurement. It may be stated in the following form:

COROLLARY 5. A physical observable A has a definite value in state Ψ if and only if Ψ is an eigenfunction of the operator A associated with A, this definite value being the corresponding eigenvalue. In symbols, this condition becomes

$$\mathsf{A}\Psi = A\Psi. \tag{4.9}$$

If Ψ_n is a solution, for which A takes the value A_n, the state represented by Ψ_n is one in which measurement of A is certain to yield the definite value A_n.

From this corollary it is evident that the separation parameter E in (4.3), which may be written $\mathsf{H}\Psi = E\Psi$ on attaching the phase factor $\exp(-iEt/\hbar)$, is not only an *expectation* value of the energy of a system in a stationary state; it is the *precise* value that will always be observed when the wave function has the form (4.2). It is also clear that in a stationary state the phase factor may be dropped: Ψ in (4.9) may then be replaced by the amplitude factor Φ, provided the operator A is time-independent. In this case (4.9) becomes

$$\mathsf{A}\Phi = A\Phi. \tag{4.10}$$

The time-independent Schrödinger equation (4.3) arises in the special case $\mathsf{A} = \mathsf{H}$.

The proof of (4.9) requires a criterion for the variable A to possess a *definite* value. If A has an average value $\bar{A} = \langle A\rangle$, calculated as in

Postulate 2, and a particular observation yields the value A, then the *deviation* is $A - \bar{A}$, and the mean square deviation is the average value of $(A - \bar{A})^2$. But the mean value of this variable, for a system in state Ψ, is obtained from its associated operator as in (4.4). The mean square deviation is thus‡

$$\Delta A^2 = \langle \Psi | (\mathsf{A} - \bar{A})^2 | \Psi \rangle. \tag{4.11}$$

The variable A may then be said to possess a definite value in state Ψ if ΔA vanishes, i.e. if there is *no deviation* from the mean, however many observations are made. The condition for a definite value is thus

$$\langle \Psi | (\mathsf{A} - \bar{A})^2 | \Psi \rangle = \langle (\mathsf{A} - \bar{A}) \Psi | (\mathsf{A} - \bar{A}) \Psi \rangle = 0$$

where the second form follows because (by Postulate 2) $(\mathsf{A} - \bar{A})$ is a Hermitian operator, and may therefore be applied indifferently to either the right-hand Ψ or the left-hand Ψ. The condition thus implies that the length of the vector $(\mathsf{A} - \bar{A}) \Psi$ must vanish, and this in turn requires (disregarding the trivial case $\Psi = 0$) that

$$(\mathsf{A} - \bar{A}) \Psi = 0.$$

This is equivalent to the eigenvalue equation (4.9) which is satisfied when $A = \bar{A} = \langle A \rangle$ is any one of the eigenvalues of A, A_n, say. Such a value is reproducible without deviation in the corresponding state Ψ_n.

EXAMPLE. *Linear and angular momentum.* We return to the Examples on p. 79 and first enquire whether the momentum p_x is *definite* in the stationary state with amplitude factor

$$\phi_k(\mathbf{r}) = L^{-3/2} \exp i(k_x x + k_y y + k_z z).$$

This being a stationary state we may use the condition in the form (4.10). Operation with $\mathsf{p}_x = (\hbar/i)\, \partial/\partial x$ yields at once

$$\mathsf{p}_x \phi_k(\mathbf{r}) = \hbar k_x \phi_k(\mathbf{r})$$

and it follows that $\hbar k_x$ is not only the mean value of p_x but is a *definite* value from which no deviations will be found in any observation on a system in state ϕ_k. On the other hand, p_x working on the real wave function (2.9a) fails to produce a multiple of ϕ and it is concluded that deviations from the average value will occur, i.e. that p_x is in this case *indefinite*.

A similar test applied to the angular momentum in the second Example shows that $L_x = L_y = L_z = 0$ in a state with a spherically symmetrical wave function. It may be verified by operating with L_z that there is definitely *one* unit of angular momentum about the z-axis in any state with a wave function of the form $(x + iy)f(r)$, while the other components are *indefinite* with zero average values.

---

‡Note that $\bar{A}$ is not a variable and therefore has no operator associated with it. It is merely a numerical constant found by averaging measured values.

4.5. The uncertainty principle

A further important consequence of Postulate 2 concerns the relationship between the probable errors (measured by the root mean square deviations) with which *two* variables, A and B say, may be measured. In its elementary form Heisenberg's uncertainty principle refers to the probable errors in measurements of the position of a free particle (x, say) and its corresponding momentum component (p_x) and asserts that

$$\Delta x\, \Delta p_x \simeq \hbar \tag{4.12}$$

—so that increasing precision in the specification of momentum of the particle implies increasing uncertainty in knowledge of its position in space. The wave packet studied in Section 1.1 illustrates this principle: Δx refers to the width of the packet (Fig. 1.1) and is proportional to $\sqrt{\sigma}$ while Δp_x refers to the spread of momentum values around the mean and is of the order $\hbar \Delta k$ or, by (1.6), $\hbar \sqrt{(1/\sigma)}$—so the product is of order $\hbar$.

The uncertainty principle may be stated most generally in the following form:

COROLLARY 6. The root mean square deviations, ΔA and ΔB, obtained by repeated measurement of observables A and B for a system in state Ψ, are related by

$$\Delta A\, \Delta B \geqq \tfrac{1}{2}\langle \Psi | \mathsf{C} | \Psi \rangle \tag{4.13}$$

where

$$i\mathsf{C} = \mathsf{AB} - \mathsf{BA} = [\mathsf{A},\, \mathsf{B}] \tag{4.14}$$

is the "commutator"‡ of the operators associated with A and B.

The proof rests upon the fact that all non-zero vectors in Hilbert space are of positive length (see (3.46) et seq.). In particular, if $\mathsf{X} = \mathsf{A} + \lambda \mathsf{B}$ (where λ is any complex number),

$$\langle \mathsf{X}\Psi | \mathsf{X}\Psi \rangle = \langle \Psi | \mathsf{A}^2 | \Psi \rangle + |\lambda|^2 \langle \Psi | \mathsf{B}^2 | \Psi \rangle$$
$$+ \lambda \langle \Psi | \mathsf{AB} | \Psi \rangle + \lambda^* \langle \Psi | \mathsf{BA} | \Psi \rangle \geqslant 0.$$

If we now let $\lambda = a + ib$, with a and b real, and introduce

$$\mathsf{AB} - \mathsf{BA} = i\mathsf{C}, \quad \mathsf{AB} + \mathsf{BA} = \mathsf{D}$$

it follows easily that

$$\langle \Psi | \mathsf{A}^2 | \Psi \rangle + (a^2 + b^2)\langle \Psi | \mathsf{B}^2 | \Psi \rangle + a\langle \Psi | \mathsf{D} | \Psi \rangle - b\langle \Psi | \mathsf{C} | \Psi \rangle \geqslant 0.$$

---

‡This definition of the commutator [A, B] is the one usually employed; some authors occasionally use this to mean $i(\mathsf{AB} - \mathsf{BA})$.

Let us take $a = 0$ and "complete the square" for the terms in b, to obtain

$$\langle\Psi|\mathsf{A}^2|\Psi\rangle + \left[b - \frac{\langle\Psi|\mathsf{C}|\Psi\rangle}{2\langle\Psi|\mathsf{B}^2|\Psi\rangle}\right]^2 \langle\Psi|\mathsf{B}^2|\Psi\rangle - \frac{\langle\Psi|\mathsf{C}|\Psi\rangle^2}{4\langle\Psi|\mathsf{B}^2|\Psi\rangle} \geqq 0.$$

Now this inequality is valid for any value of b, which we may therefore choose so that the second term vanishes.‡ The result is

$$\langle\Psi|\mathsf{A}^2|\Psi\rangle \, \langle\Psi|\mathsf{B}^2|\Psi\rangle \geqq \tfrac{1}{4}\langle\Psi|\mathsf{C}|\Psi\rangle^2.$$

To introduce the deviations we denote the expectation values by $\bar{A}$ and $\bar{B}$ and note that, as in (4.11),

$$\Delta A^2 = \langle\Psi|(\mathsf{A}-\bar{A})^2|\Psi\rangle \quad \Delta B^2 = \langle\Psi|(\mathsf{B}-\bar{B})^2|\Psi\rangle.$$

It is also clear that $(\mathsf{A}-\bar{A})$ and $(\mathsf{B}-\bar{B})$ have the same commutator, namely $i\mathsf{C}$, as A and B themselves. On inserting the former operators in the left-hand side of the inequality, instead of A and B, and using the definitions of ΔA and ΔB, it follows at once that

$$\Delta A^2 \times \Delta B^2 \geqq \tfrac{1}{4}\langle\Psi|\mathsf{C}|\Psi\rangle^2$$

which establishes the desired result (4.13).

EXAMPLE. *Uncertainties in x and p_x.* In Schrödinger language x is simply a multiplier, while p_x is the differential operator $(\hbar/i)\,\partial/\partial x$. The commutator thus has the effect, on any operand Ψ

$$i\mathsf{C}\Psi = -i\hbar\left(x\frac{\partial\Psi}{\partial x} - \frac{\partial}{\partial x}(x\Psi)\right).$$

The second term is then differentiated as a product and we obtain, discarding the arbitrary operand,

$$i\mathsf{C} = \mathsf{x}\mathsf{p}_x - \mathsf{p}_x\mathsf{x} = i\hbar\mathsf{I}$$

where I is the unit operator which leaves any Ψ unchanged. In this case, then, (4.13) reduces to

$$\Delta x \Delta p_x \geqq \tfrac{1}{2}\hbar$$

which fixes a precise lower bound in (4.12).

Finally, (4.13) shows that *if the operators A and B commute*, the lower bound on $\Delta A \Delta B$ for the quantities with which they are associated is zero; and thus ΔA, or ΔB, or *both* may vanish. This latter possibility would suggest the existence of states in which *two* variables could simultaneously be assigned perfectly definite values. This is reminiscent of the classical situation in which, besides the energy, there may be additional "constants of the motion". In the motion of a planet, for example, the energy and angular momentum components (and hence

---

‡Note that C is Hermitian so that $\langle\Psi|\mathsf{C}|\Psi\rangle$ is real and is thus compatible with the assumption that b is a real number.

the plane of the orbit) all have definite and unchanging values for any given state of motion. We consider this situation in more detail in a later section.

4.6. Time-development and the energy-time uncertainty principle

From Postulates 1 and 2 it is possible to derive a general expression for the rate of change of the expectation value of any quantity A, with operator A. Thus, from (4.4),

$$\frac{d}{dt}\langle A\rangle = \langle(\partial\Psi/\partial t)|A|\Psi\rangle + \langle\Psi|A|(\partial\Psi/\partial t)\rangle + \langle\Psi|(\partial A/\partial t)|\Psi\rangle$$

$$= \langle A\Psi|(\partial\Psi/\partial t)\rangle^* + \langle\Psi|A|(\partial\Psi/\partial t)\rangle + \langle\Psi|(\partial A/\partial t)|\Psi\rangle.$$

But from (4.1) this becomes

$$\frac{d}{dt}\langle A\rangle = \left[\frac{1}{i\hbar}\langle A\Psi|H\Psi\rangle\right]^* + \frac{1}{i\hbar}\langle\Psi|AH\Psi\rangle + \langle\Psi|(\partial A/\partial t)|\Psi\rangle$$

and since, using the Hermitian symmetry of the operator, $\langle A\Psi|H\Psi\rangle = \langle\Psi|AH\Psi\rangle$ we obtain finally

---

COROLLARY 7. The time rate of change of any expectation value is given by

$$i\hbar\frac{d}{dt}\langle A\rangle = \langle\Psi|(AH-HA)|\Psi\rangle + \langle\Psi|(\partial A/\partial t)|\Psi\rangle \qquad (4.15)$$

where the system is described by the state vector Ψ, whose time development is determined by Postulate 1.

This result gives a special significance to the commutator of an arbitrary operator and the *Hamiltonian*: if the operator does not depend explicitly on the time, and if it commutes with the Hamiltonian, then its expectation value will also be independent of the time. In other words, a quantity whose associated operator has these properties will be a *constant of the motion*. The simplest example of such an operator is the Hamiltonian of a conservative dynamical system $(\partial H/\partial t = 0)$: the energy $E = \langle H\rangle$ is then obviously a constant of the motion and we obtain the quantum analogue of the classical principle of energy conservation. It should be noted particularly that this result is valid even when the system is not in a stationary state and is therefore not entirely trivial; it applies, for example, to the motion of a wave packet in which the energy is not precise (corresponding to a range of k values in the example on p. 35)—the *expectation* energy remains constant so long as the system is not disturbed by, for example, time-dependent fields.

Now that we know how an expectation value develops in time, it is possible to consider an uncertainty principle connecting energy and time, usually stated in the form

$$\Delta E \, \Delta t \simeq \hbar. \tag{4.16}$$

This result is often applied, somewhat naïvely, in discussing phenomena ranging from the observation of free particles to the lifetimes of stationary states of complicated systems. Its status is, however, quite different from that of (4.13); the meaning of Δt, for example, needs very careful definition because t is not a dynamical observable but rather a numerical parameter describing the evolution of the system. To obtain an uncertainty relation involving E we may start from (4.13), assuming A is not explicitly dependent on time and taking $\mathsf{B} = \mathsf{H}$: the result is

$$\Delta E \, \Delta A \geqq \tfrac{1}{2} \langle \Psi | (1/i((\mathsf{AH}-\mathsf{HA}) | \Psi \rangle.$$

But from the corollary just established the right-hand side of this equation measures the time rate of change of $\langle A \rangle$. Substitution of (4.15) yields

$$\Delta E \, \Delta t \geqq \tfrac{1}{2}\hbar \tag{4.17}$$

where we have *defined* Δt as the quantity (which certainly has dimensions of time)

$$\Delta t = \frac{\Delta A}{d/dt \langle A \rangle}. \tag{4.18}$$

The uncertainty relation (4.17) is now quite precise and is formally similar to that for position and momentum (p. 83); it remains only to clarify the meaning of Δt as defined in (4.18). Since Δt times the rate of change of $\langle A \rangle$ indicates the amount by which the expectation value of a typical dynamical quantity would change during time Δt, we may express the above results verbally in the form:

COROLLARY 8. There is an energy–time uncertainty relation

$$\Delta E \, \Delta t \geqslant \tfrac{1}{2}\hbar$$

where ΔE is the (root-mean-square) uncertainty in the energy of the system and Δt is the time needed for the average value of a dynamical variable A to change by an amount comparable to its uncertainty ΔA and is defined precisely by (4.18).

An example clarifies the meaning of Δt. If a system initially in a stationary state is put in interaction with its surroundings (i.e. the

Hamiltonian is modified) it will not remain in that stationary state and there will be a corresponding uncertainty $\Delta E \neq 0$: the state will evolve in time and the change of expectation value of any quantity A will become detectable (i.e. comparable in magnitude with the intrinsic uncertainty ΔA) after a time of order not less than $\frac{1}{2}\hbar/\Delta E$. Thus the "lifetime" of state of *precise* energy ($\Delta E = 0$) is infinite; but strong interaction (large ΔE) will be related to short lifetime.

4.7. The completeness of eigenfunction sets

In the Schrödinger representation, the state of a particle is described by the wave function $\psi(\mathbf{r};t)$ and we have already met situations in which this function is written as a linear combination of the particular wave functions, eigenfunctions, characterizing the stationary states of the system. Thus, a particle in a box with periodic boundary conditions (p. 36) might be described by the wave packet

$$\psi(\mathbf{r};t) = \sum_k c_k \psi_k(\mathbf{r};t) \tag{4.19}$$

where the $\psi_k(\mathbf{r};t)$ are plane wave solutions corresponding to states of definite momentum $p = \hbar k$. Since $\psi_k(\mathbf{r};t) = \phi_k(\mathbf{r}) \exp(-iE_k t/\hbar)$, the expansion may be written alternatively

$$\psi(\mathbf{r};t) = \sum_k c_k(t) \phi_k(\mathbf{r}) \tag{4.20}$$

where the coefficients are time-dependent and the ϕ_k are the eigenfunctions of H; satisfying $H\phi_k = E_k\phi_k$. Equation (4.20) is an example of an *eigenfunction expansion*; the basic theory underlying such expansions has been covered in Sections 3.2 and 3.3.

In some cases the summation index in an expansion such as (4.20) may assume continuous values (as, for example, with a free particle, in which all values of the momentum are allowed) and the sum must then be replaced by an integral. For ease of exposition and notation, however, we shall normally assume that the eigenvalues form a *discrete* set. The assumption we now wish to make is that *any state* of a system may be expressed as a linear combination of its eigenstates. We express this idea as a postulate:

POSTULATE 3. The solutions of the eigenvalue problem

$$H\Phi = E\Phi$$

for a system with Hamiltonian operator H, constitute a complete set, closed under the action of all the operators of the system.

The completeness of the set means simply that any state vector may be expanded in the form

$$\Psi = \sum_k a_k \Phi_k, \tag{4.21}$$

where the a_k are numerical coefficients (which in general may depend on the time). The property of "closure" means that if A is any dynamical operator of the system then $A\Psi$ can be expanded in terms of the same set:

$$A\Psi = \sum_k b_k \Phi_k. \tag{4.22}$$

When the Φ_k are regarded as basis vectors of a Hilbert space, such expansions have the geometrical interpretation discussed in Section 3.5.

In Schrödinger language, Ψ and the Φ's are functions of class Q (p. 18) and H is a differential operator; examples of sets we have already met are the eigenfunctions of the one-dimensional box (Section 2.1) which are complete for all functions $\psi(x)$ defined in the interval $(0, L)$; those of the three-dimensional box (Section 2.1), complete for functions $\psi(x, y, z)$ defined within a cube of side L; those of the one-dimensional harmonic oscillator (Section 2.2) complete for all functions $\psi(x)$ defined in the interval $(-\infty, +\infty)$. Other sets exist which are complete for all functions (of class Q) defined in the whole of three-dimensional space. Such sets are all infinite and the number of significant terms in an expansion of the form (4.21) will of course depend on the nature of ψ; if ψ is quite close to one of the eigenfunctions, ϕ_k, a good representation may be obtained with few terms. The postulate refers, in general, to a many-particle system.

Unfortunately, the eigenfunctions of interest very often cannot be expressed in closed form. It is for such reasons that we are compelled to assume completeness as a postulate.

It should be noted that Postulate 3 refers to state vectors and eigenstates "of a given system"; complete sets are frequently used in a wider context, without reference to a particular eigenvalue problem, but naturally for an expansion to be valid all the Φ_k must at least be functions of the same class. As an example of this wider usage we note that the hydrogen atom eigenfunctions may be expressed with arbitrarily high accuracy in terms of those of a three-dimensional harmonic oscillator by taking a sufficient number of terms; Ψ and the Φ_k then refer to different physical systems, but the expansion is still valid because the Φ_k are complete for all functions of position in three-dimensional space. On the other hand, the eigenfunctions of a finite box could not be used for this purpose since they are complete only for the region within the box.

Returning to the Postulates, we note first an immediate consequence of the Hermitian symmetry of H:

COROLLARY 9. The solutions Φ_k of the eigenvalue equation

$$\mathsf{H}\Phi = E\Phi$$

may be chosen so as to form an orthonormal set:

$$\langle \Phi_i | \Phi_j \rangle = \delta_{ij}.$$

This is essentially a mathematical property of Hermitian operators, the proof following closely that given in Section 3.1 (for differential operators and their eigenfunctions) and referred to again in Sections 3.3 and 3.7. We remember that orthogonality is "automatic" for eigenfunctions with different eigenvalues, and may be assumed without loss of generality (i.e. orthogonal solutions *can* be found) when degeneracy occurs. The possibility of normalizing the solutions follows from Corollary 1. When degeneracy does occur, it indicates that measurement of E is not sufficient to characterize a state completely; Φ_i and Φ_j may have a common energy $E_i = E_j$ but differ in the values taken by some other observable A. To investigate this possibility we first use Postulate 3 to verify the conjecture at the end of Section 4.5. The basic result needed may be expressed in the form

---

COROLLARY 10. Two operators A and B possess a complete set of common eigenvectors if and only if they commute:

$$[\mathsf{A}, \mathsf{B}] = \mathsf{AB} - \mathsf{BA} = 0.$$

The eigenvectors then describe states in which variables A and B have simultaneously definite values.

The necessity of this condition is easily demonstrated: for if Ψ_k is a common eigenvector‡ of A and B, with eigenvalues A and B, then

$$\mathsf{AB}\Psi_k = \mathsf{A}(B\Psi_k) = B\mathsf{A}\Psi_k = AB\Psi_k,$$

$$\mathsf{BA}\Psi_k = \mathsf{B}(A\Psi_k) = A\mathsf{B}\Psi_k = AB\Psi_k,$$

and hence

$$(\mathsf{AB} - \mathsf{BA})\Psi_k = 0.$$

---

‡And hence some vector in the Hilbert space of the system considered, which is assumed closed under A and B.

If this is true for every member of a complete set we may multiply by c_k and sum over all k to obtain

$$(AB - BA)\Psi = 0$$

where Ψ is the arbitrary function

$$\Psi = \sum_k c_k \Psi_k.$$

But this is true for arbitrary Ψ only if $(AB - BA)$ is the zero operator and the condition is thus established.

The sufficiency of the condition, allowing us to assert that if A and B commute then a complete set can be found, such that all its members are simultaneously eigenfunctions, is less easily proved.‡ The proof will only be sketched. First, it can be shown that a basis $\{\Phi_k\}$ may be chosen, without any loss of generality, so that its members satisfy §

$$A\Phi_k = A_k \Phi_k, \tag{4.23}$$

i.e. that using *any* complete set for the system, its members may be formed into new linear combinations that are eigenvectors of A. This result depends only on the property of closure, together with the Hermitian symmetry of the operator; for a *finite* vector space the proof is essentially contained in Section 3.7, but more generally we accept the statement as a purely mathematical result (see, for example, Courant and Hilbert (1953)). The proof of sufficiency then runs as follows. If B commutes with A, and the Φ_k are eigenfunctions of A, then (using Hermitian symmetry)

$$\langle \Phi_i | (AB - BA) | \Phi_j \rangle = (A_i - A_j)\langle \Phi_i | B | \Phi_j \rangle = 0$$

and hence $\langle \Phi_i | B | \Phi_j \rangle = 0$ provided $A_i \neq A_j$. But, by the closure property, $B\Phi_j$ can be expanded in the form

$$B\Phi_j = \sum_k b_{kj} \Phi_k, \tag{4.24}$$

where the coefficients follow on taking a scalar product with Φ_k:

$$b_{kj} = \langle \Phi_k | B | \Phi_j \rangle \tag{4.25}$$

---

‡A careful discussion by Kemble (*loc. cit.*, pp. 181–5) indicates the need for further assumptions before the usual "proofs" can be justified; these amount essentially to the closure property adopted in Postulate 3.

§Note that Φ_k is here not necessarily an eigenvector of the *Hamiltonian* operator. Nor is any assumption made about time dependence; time-dependent vectors are expressible in a time-independent basis, the expansion coefficients being time-dependent.

This coefficient is non-zero only for $k = j$, as noted above, provided the eigenvalues of A are non-degenerate and therefore in this case (on using (4.25) in (4.24) with $k = j$)

$$B\Phi_j = b_{jj}\Phi_j = B_j\Phi_j \qquad (4.26)$$

showing that the eigenfunctions of A are simultaneously eigenfunctions of B. If an m-fold degenerate eigenvalue occurs we write a *general* eigenvector, with that eigenvalue, as a linear combination of the m independent solutions; it can then be shown (cf. the Example on p. 47) that the coefficients can be chosen so as to define m eigenvectors of A that *are*, simultaneously, eigenvectors of B.

We may now return to the use of commuting operators as a means of classifying stationary states of a system in terms of constants of the motion—one of their most important applications. In this case one of the commuting operators is the Hamiltonian itself. If another operator A commutes with H, then we can find states in which A has a definite value along with the energy. If we can find a third operator B, which also commutes with H *and with* A, it follows that we can find a complete set of states in which the *three* quantities, E, A, B, have simultaneously definite values—and so on. The simultaneous eigenvalues may thus be used to give an increasingly precise classification of the state of the system; and the largest set of mutually commuting operators that we can find will give the most complete characterization possible. It follows from Corollary 7 that the simultaneous eigenvalues are all constants of the motion in the sense that they are perfectly definite and unchanging. No more complete specification of the state can be obtained; it is a "state of maximal knowledge".

States of maximal knowledge are of fundamental importance in the theory of measurement, representing the ultimate limits of precise observation permitted by the uncertainty principle. Any attempt to measure another variable (with a *non*-commuting operator) must then introduce uncertainty into at least one of those already measured.‡ A "sharper" specification of the state is therefore unattainable. Observables whose operators all commute with each other, and which are therefore simultaneously knowable, are said to be *compatible*. The simplest illustration of these important concepts is afforded by a system consisting of a single free particle (Section 2.4):

---

‡Any effective measurement of the new variable must leave it with a *finite* (or zero) un-certainty; the uncertainty introduced into one of the measured variables must therefore be non-zero, according to Corollary 6, if the corresponding commutator is non-zero.

EXAMPLE. *Commuting operators for a free particle.* It is clear by inspection that the operators (using, as usual, the Schrödinger language)

$$= -\frac{\hbar^2}{2m}\frac{\partial^2}{\partial x^2}, \quad \mathsf{P}_x = \frac{\hbar}{i}\frac{\partial}{\partial x}, \quad \mathsf{P}_y = \frac{\hbar}{i}\frac{\partial}{\partial y}, \quad \mathsf{P}_z = \frac{\hbar}{i}\frac{\partial}{\partial z}$$

all commute with each other; $\mathsf{P}_x\mathsf{P}_y = \mathsf{P}_y\mathsf{P}_x$, for example, because the order of successive differentiations is immaterial for any well-behaved function. We have also verified directly (p. 81) that the wave function

$$\phi_k(\mathbf{r}) = L^{-3/2}\exp(i\mathbf{k}\cdot\mathbf{r})$$

is an eigenfunction of all four operators, with eigenvalues

$$E = \hbar^2 k^2/2m, \quad p_x = \hbar k_x, \quad p_y = \hbar k_y, \quad p_z = \hbar k_z$$

respectively. This is a state of maximal knowledge for a free particle, the energy and momentum components all being constants of the motion.

Finally, having introduced the idea of states of maximal knowledge, we may formulate one more important corollary:

COROLLARY 11. If the state vector Ψ of a system is expressed in terms of the eigenvectors Ψ_k describing states of maximal knowledge,

$$\Psi = \sum c_k\Psi_k, \tag{4.27}$$

then the probability that, in an experiment designed to yield maximal knowledge, the system will be found in state Ψ_k is

$$w_k = |c_k|^2 = |\langle\Psi_k|\Psi\rangle|^2. \tag{4.28}$$

The scalar products $\langle\Psi_k|\Psi\rangle$ in this way acquire a physical meaning.

This interpretation follows on examining the expectation values, in state Ψ, of all the compatible variables $E, A, B, C, \ldots$ whose eigenvalues characterize the states. Repeated measurements, always starting with the system in state Ψ, would yield an average value of A given by

$$\langle A\rangle = \langle\Psi|\mathsf{A}|\Psi\rangle = \sum_{k,l} c_k{}^* c_l\langle\Psi_k|\mathsf{A}|\Psi_l\rangle.$$

But since Ψ_k, Ψ_l are not only orthogonal but are also simultaneous eigenfunctions of $\mathsf{H}, \mathsf{A}, \mathsf{B}, \mathsf{C}, \ldots$, we obtain

$$\langle A\rangle = \sum_{k,l} c_k{}^* c_l A_l\langle\Psi_k|\Psi_l\rangle = \sum_k |c_k|^2 A_k$$

and similarly for B, C, ... Thus, with $w_k = |c_k|^2$

$$\langle A \rangle = \sum w_k A_k, \quad \langle B \rangle = \sum w_k B_k, \quad \ldots \qquad (4.29)$$

These average values indicate that the relative frequency with which the system is found in state Ψ_k, with eigenvalues E_k, A_k, B_k, . . . is w_k, and the corollary is established.

A more complete discussion of this result and its implications would require a detailed discussion of the nature of physical measurements, and lies outside the scope of this book. We note, however, that when a system in state Ψ is subjected to observations, leading to definite knowledge of a maximal set of observables, it makes a "transition" from state Ψ to some state Ψ_k corresponding to maximal information *about the state in which the system is left.* In this idealized process the system is "forced" into one of the states of maximal knowledge, as a result of interaction with the measuring apparatus; during the process, whose end result is $\Psi \rightarrow \Psi_k$, the state does not evolve according to the normal causal law—which refers to the system itself, unconnected with measuring apparatus. Thus, w_k in Corollary 11 indicates the probability that measurement of the maximal set of compatible variables (as chosen in defining the expansion) will leave the system in the particular state Ψ_k, without reference to *how* the change is effected. It should be noted that this interpretation of the coefficients in Corollary 11 is dependent upon the Ψ_k being eigenfunctions *of the system to which Ψ refers.* It is, for example, entirely possible for the eigenfunctions of the electron in a hydrogen atom to be expanded in a complete set of harmonic oscillator functions; but w_k would clearly *not* represent the probability of finding the electron in a harmonic oscillator state. The squared coefficients in a general expansion, with only mathematical significance, are frequently referred to as "probabilities"; it is safer to use the term "weights" unless the eigenfunctions Ψ_k actually refer to the system in question.

4.8. Properties of the operators

The operators associated with physical quantities have so far been specified only in the Schrödinger language, by means of the rule (4.6). The postulates and their consequences, on the other hand, have been formulated in a much more general manner, not depending on this somewhat mysterious rule. To complete the general formulation of quantum mechanics we therefore need to define the operators themselves in a language-independent manner. The natural way of doing this is by specifying the rules by which they combine with each other. The fundamental significance of the *commutator* of two operators has already been established, and it is therefore not surprising that the operator algebra of quantum mechanics is completely defined by a set of *commutation rules*. These rules may now be stated in the form of another postulate:

POSTULATE 4. The operators associated with the position and momentum variables of a particle commute, *with the following exceptions*

$$[\mathsf{x}, \mathsf{p}_x] = [\mathsf{y}, \mathsf{p}_y] = [\mathsf{z}, \mathsf{p}_z] = i\hbar\mathsf{l} \qquad (4.30)$$

while those for two *different* particles always commute.

Such properties have already been noted (e.g. in the Example on p. 83) in the case of position and momentum operators in Schrödinger language, set up using the prescription (4.6) et seq. The statement now made is more general in the sense that x, p_x, etc., are simply Hermitian operators, and the operand Ψ is simply a vector in a Hilbert space—not necessarily a function $\Psi(\mathbf{x})$. Such operators may be specified in an infinite variety of ways, by noting their effect on *any* complete set Φ_1, Φ_2, ... defining the space in which they operate; but the algebraic relationships among them are independent of the language or *representation* chosen.

Before considering other quantum mechanical languages (Chapter 5) we note that the form of Postulate 4 is largely prescribed (see Temple, 1934, pp. 44–46) by the more general principle that space is *isotropic* (i.e. the x, y, z directions are arbitrary); and, secondly, that the association of operators with various *functions* of the basic variables (position and momentum) must be compatible with Postulate 3. Thus, for example, the expectation value of $A + B$ is

$$\langle (A+B) \rangle = \langle A \rangle + \langle B \rangle$$

by definition of the expectation value as an arithmetic mean.‡ Consequently remembering the operators are assumed linear,

$$\langle (A+B) \rangle = \langle \Psi|\mathsf{A}|\Psi \rangle + \langle \Psi|\mathsf{B}|\Psi \rangle = \langle \Psi|(\mathsf{A}+\mathsf{B})|\Psi \rangle \quad (\text{all } \Psi)$$

and the operator to be associated with $A + B$ must therefore be $\mathsf{A} + \mathsf{B}$.

---

‡The meaning should be considered carefully. Each variable is measured a large number of times (N, say), independently and always starting with the system in state Ψ: a pair of measurements, A_i and B_i give one estimate of the sum $A + B$ and the arithmetic mean $N^{-1} \sum_i (A_i + B_i)$ is the sum of the arithmetic means of A and B separately. The question of whether the measurements are compatible or incompatible does not arise.

Similar arguments (Temple, *loc. cit.*, pp. 36–38) lead to the following requirements:

COROLLARY 12. The operators associated with multiples, sums and products of observables are as follows:

$$cA \rightarrow c\mathsf{A}$$

$$A + B \rightarrow \mathsf{A} + \mathsf{B}$$

$$AB \rightarrow \tfrac{1}{2}(\mathsf{AB} + \mathsf{BA})$$

where the $\rightarrow$ means the operator on the right is associated with the observable on the left.

Since any dynamical quantity can be expressed in terms of sums and products of the basic variables x, y, z, p_x, p_y, p_z, the associated operator may be set up using Corollary 12: the properties of such operators are then determined by the basic rules of Postulate 4. Thus, for example, the operators associated with xp_y and yp_x are $\mathsf{x}\mathsf{p}_y$ and $\mathsf{y}\mathsf{p}_x$ (since in this case the operators commute); that associated with $-yp_x$ is $-\mathsf{y}\mathsf{p}_x$; and hence the operator associated with the angular momentum $L_z = xp_y - yp_x$ is $\mathsf{L}_z = \mathsf{x}\mathsf{p}_y - \mathsf{y}\mathsf{p}_x$. Generally, any function of x, y, $\ldots$ p_z gives rise to an operator which is a similar function (suitably symmetrized in the rare cases where it may be necessary) of the operators x, y, $\ldots$, p_z. Corollary 12 merely justifies formally the rules previously adopted.

In conclusion, it must be added that a discussion of the more mathematical properties of the operators, in particular questions of "boundedness" and the nature of the eigenvalue spectrum, require sophisticated analysis. Such analysis is clearly outside the scope of this book but may be found elsewhere (e.g. Courant and Hilbert, 1953; von Neumann, 1955; Kato, 1951).

4.9. Electron spin

So far, although the postulates have been framed in a general way, it has been assumed that all the operators encountered in the theory may be derived from classically defined dynamical quantities (i.e. functions of particle coordinates and momenta); and we know how to set up these operators using the Schrödinger language. Experiment shows, however, that there are some quantities that have no precise classical analogues; to complete the postulates we must introduce these quantities and the properties that characterize their associated operators.

We shall not give a detailed historical account of the development of the "spin" concept; it is sufficient to start from the experimentally established premise that a particle may exhibit properties that would correspond, in classical terms, to an *intrinsic angular momentum* and a related *magnetic moment*. For example, the state of a single electron moving in an arbitrary potential field is always (at least) doubly degenerate, application of a magnetic field resolving even an apparently non-degenerate energy level into two branches, with a minute separation proportional to the applied field strength; these branches evidently correspond to some observable that can take only two values. If the magnetic moment were represented by a vector μ, the classical interaction energy with an applied magnetic field $\mathbf{B}$ would be $-\mathbf{B} \cdot \mu$ or, if the field were adopted as the z-axis, $-B\mu_z$ where μ_z is the component of the dipole vector along the field direction; and if μ were taken proportional to a "spin angular momentum" the classically expected interaction energy would be proportional to $-BS_z$. The experiments imply that, for an electron, the z-component of spin S_z can take only two values; this situation is described by saying that the spin vector sets parallel or antiparallel to an applied field.

Let us now introduce the spin formally and quantitatively by means of the following Postulate:

POSTULATE 5. An electron possesses an intrinsic angular momentum‡ represented by a "spin" vector $\mathbf{S}$ with components S_x, S_y, S_z, each being a "two-valued" observable, with possible values $\pm\frac{1}{2}$. The related magnetic moment is $\mu = -g\beta\mathbf{S}$ where $-\beta$ is the classically expected moment for one atomic unit of angular momentum, and $g = 2\cdot0023$ is an observed "free-electron g value". The associated spin *operators* commute with all the operators representing "classical" quantities, but not with each other.

The classically expected magnetic moment, for an electron with 1 atomic unit of *orbital* angular momentum is $-e\hbar/2m$ and the "Bohr magneton" is defined by§

$$\beta = e\hbar/2m \qquad (4.31)$$

---

‡As on p. 79 we measure angular momenta in units of $\hbar$, i.e. in atomic units. Thus $\hbar S_x$ is the angular momentum component corresponding to a dimensionless S_x.

§Here we use SI units, postponing a full discussion to Vol. 2, where electric and magnetic effects are considered in more detail. In the mixed Gaussian system, β contains an additional factor c in the denominator.

The minus sign in $\mu = -g\beta S$ arises from the negative charge of the electron. The fact that $g \simeq 2$ for *spin* angular momentum is classically unexpected, but emerges naturally from the Dirac equation (Chap. 6, Vol. 2) which takes account of the requirements of relativity theory; the precise value $2 \cdot 0023$ is predicted when further refinements are admitted. In non-relativistic quantum mechanics, electron spin is admitted essentially by Postulate 5, which forms the basis of the *Pauli theory* of spin.

In order to assimilate spin into the theory we note first that measurement of any spin component, that along the z-axis say, yields one of two possible eigenvalues $+\frac{1}{2}\hbar$ or $-\frac{1}{2}\hbar$; we denote these corresponding eigenstates by α and β and write

$$\mathsf{S}_z \alpha = \tfrac{1}{2}\alpha, \qquad \mathsf{S}_z \beta = -\tfrac{1}{2}\beta. \tag{4.32}$$

It is then possible to describe simultaneous eigenstates, in which the z-component of spin is definite along with various other "classical" quantities (e.g. energy, momentum), as *products* of an orbital factor ϕ and a spin factor, α or β; spin operators work only on the latter, all other operators only on the orbital factor ϕ. Thus, if an electron in a beam has linear momenta corresponding to wave numbers k_x, k_y, k_z (p. 36), so that

$$\mathsf{p}_x \phi = (k_x \hbar)\phi, \quad \mathsf{p}_y \phi = (k_y \hbar)\phi, \quad \mathsf{p}_z \phi = (k_z \hbar)\phi$$

then the state is more completely characterized as $\phi\alpha$ or $\phi\beta$. The spin component along the (arbitrarily chosen) z-axis being $\pm\frac{1}{2}\hbar$ in the two cases. Thus, for example,

$$\mathsf{S}_z(\phi\alpha) = \phi(\mathsf{S}_z\alpha) = \phi(\tfrac{1}{2}\alpha) = \tfrac{1}{2}(\phi\alpha)$$

showing that the *product* is also a spin eigenfunction. Also, $\phi\alpha$ clearly remains an eigenfunction of the other operators p_x, p_y, p_z, which work only on the ϕ-factor. As a result of this separability of space and spin descriptions, we may now turn attention directly to the spin operators S_x, S_y and S_z and their eigenstates. The small magnetic interactions between spin and orbital motion will be discussed elsewhere (Vol. 2).

Commutation relations

The *non*-commutation of different spin operators is an expression of the experimental fact that *two* or more components cannot take simultaneously definite values (Corollary 10). If we start from a state with spin factor α and then measure S_x we may find $S_x = \pm\frac{1}{2}$; on repeating the measurement of S_z, however, we no longer find exclusively $S_z = +\frac{1}{2}$—either value is equally likely, giving a zero average value, and we have thus lost all knowledge of S_z. This is an example of the incompatibility of observables with non-commuting operators. We now enquire whether there are any precise commutation rules for the spin operators, parallel to those in Postulate 4.

We select one direction in space, calling this the z-axis, and try to describe any spin state as a linear combination of the eigenstates α and β satisfying (4.32). In accordance with the general principles, we regard α and β as orthogonal unit vectors defining a two-dimensional *spin space*, in which

$$\langle \alpha | \alpha \rangle = \langle \beta | \beta \rangle = 1, \quad \langle \alpha | \beta \rangle = \langle \beta | \alpha \rangle^* = 0. \tag{4.33}$$

A general state, for example one with spin component $+\frac{1}{2}$ along a direction *perpendicular* to the z-axis, might then be represented by $\eta = a\alpha + b\beta$, where the unit vector condition requires $|a|^2 + |b|^2 = 1$.‡ Similarly, we try to characterize all three spin operators S_x, S_y, S_z in terms of their effects on the basis vectors α and β.

The starting-point in the argument is simply that (except in general relativity theory) all directions in space are considered equivalent; if S_z' is the operator corresponding to measurement of spin component along any new, rotated z-axis, then S_z' must have properties exactly like S_z—in particular, its eigenvalues must be $\pm\frac{1}{2}$. In general (Section 3.6) the components of a vector relative to a rotated coordinate frame are connected with those for a "fixed" frame by equations such as $S_z' = lS_x + mS_y + nS_z$ where l, m, n are direction cosines of the rotated z-axis relative to the fixed axes§ and therefore (Corollary 12) the *operator* associated with measurement of spin component along a new z-axis will be

$$S_z' = lS_x + mS_y + nS_z. \tag{4.34}$$

In particular, S_x, S_y and S_z must have similar properties. Now S_z^2 is seen to multiply α and β—and hence any linear combination—by $\frac{1}{4}$, and therefore the square of *each* spin operator is $\frac{1}{4}$ times the unit operator (I say):

$$S_x^2 = S_y^2 = S_z^2 = \tfrac{1}{4}I. \tag{4.35}$$

An immediate result is that in *any* spin state η the square of the total spin has the eigenvalue $\frac{3}{4}$: in other words

$$\mathbf{S}^2\eta = (S_x^2 + S_y^2 + S_z^2)\eta = \tfrac{3}{4}\eta. \tag{4.36}$$

---

‡Experiment shows, in this case, that $|a| = |b|$ in order to yield a zero expectation value of S_z: for $\langle S_z \rangle = \langle \eta | S_z | \eta \rangle = \frac{1}{2}(|a|^2 - |b|^2)$ and the vanishing of this quantity implies equal probabilities of the two possible components along the z-axis. Thus $\eta = (\alpha + \beta)/\sqrt{2}$ is one example of a spin state in which the z-component is indefinite but the component in a perpendicular direction *may* be definite.

§The rotation of axes is here described by the matrix $\mathbf{T}(=\mathbf{R})$ in the Example on p. 62. The components of a fixed vector change according to (3.61) where, since $\mathbf{T}$ is a real orthogonal matrix, $\mathbf{T}^{-1} = \mathbf{T}^\dagger = \tilde{\mathbf{T}}$ (the transpose).

We usually write the eigenvalue of $\mathbf{S}^2$ in the form $S(S+1)$, which will turn out to be characteristic of angular momenta in general, and when $S = \frac{1}{2}$ we refer to a "spin $\frac{1}{2}$ particle". Thus electrons are particles of spin $\frac{1}{2}$—but this means that the observed magnitude of the spin (square root of the observable value of $\mathbf{S}^2$) is $\sqrt{\{\frac{1}{2}(1+\frac{1}{2})\}}$, a result first inferred experimentally in the early days of quantum theory.

The next step is to consider $S_z'^2 = \frac{1}{4}\mathsf{I}$ where S_z' is given by the general expression (4.34). Clearly

$$S_z'^2 = (l^2 S_x^2 + m^2 S_y^2 + n^2 S_z^2) + lm(S_x S_y + S_y S_x) + mn(S_y S_z + S_z S_y)$$
$$+ nl(S_z S_x + S_x S_z).$$

From (4.35), and the fact that $l^2 + m^2 + n^2 = 1$, the first parentheses yields $\frac{1}{4}\mathsf{I}$; and since the whole expression must be equivalent to $\frac{1}{4}\mathsf{I}$, for all values of l, m, n, it follows that the spin operators must *anticommute*

$$(S_x S_y + S_y S_x) = (S_y S_z + S_z S_y) = (S_z S_x + S_x S_z) = 0 \qquad (4.37)$$

where 0 denotes the zero operator.

To find the commutators we use simple algebraic arguments. First, the condition $(S_x S_z + S_z S_x)\alpha = 0$ yields, using (4.32),

$$S_z(S_x\alpha) = -\tfrac{1}{2}(S_x\alpha)$$

and hence $(S_x\alpha)$ with eigenvalue $-\frac{1}{2}$, must be a multiple of β. Similarly, $(S_x\beta)$ is a multiple of α, and we therefore write

$$S_x\alpha = k\beta, \quad S_x\beta = k'\alpha.$$

Since S_x has real eigenvalues it is a Hermitian operator and hence

$$\langle\alpha|S_x\beta\rangle = \langle S_x\alpha|\beta\rangle$$

and this shows $k' = k^*$. Also $S_x^2\alpha = kS_x\beta = kk'\alpha$ shows that $kk' = \frac{1}{4}$. Hence

$$k = \tfrac{1}{2}e^{i\theta}, \quad k' = \tfrac{1}{2}e^{-i\theta}.$$

We take $\theta = 0$ and write

$$S_x\alpha = \tfrac{1}{2}\beta, \quad S_x\beta = \tfrac{1}{2}\alpha \qquad (4.38)$$

the choice being open because each basis vector is arbitrary to within a phase factor (Corollary 1).

A similar argument, based on the second anti-commutator in (4.37), yields

$$S_y\alpha = k\beta \qquad S_y\beta = k'\alpha$$

where k and k' are again conjugate complex numbers of modulus $\frac{1}{2}$. But use of the first anticommutation condition then shows

$$(S_x S_y + S_y S_x)\alpha = S_x(k\beta) + S_y(\tfrac{1}{2}\beta) = \tfrac{1}{2}k\alpha + \tfrac{1}{2}k'\alpha = 0$$

and hence $k = -k' = -k^*$. Therefore k is pure imaginary and must be $\pm\frac{1}{2}i$; we take

$$S_y\alpha = \tfrac{1}{2}i\beta, \quad S_y\beta = \tfrac{1}{2}i\alpha \tag{4.39}$$

choosing the upper sign for reasons that will be clear presently.

Finally we verify that

$$(S_x S_y - S_y S_x)\alpha = S_x(\tfrac{1}{2}i\beta) - S_y(\tfrac{1}{2}\beta) \;=\; \tfrac{1}{2}i(\tfrac{1}{2}\alpha) - \tfrac{1}{2}(-\tfrac{1}{2}i\alpha) = \tfrac{1}{2}i\alpha \;= iS_z\alpha,$$

$$(S_x S_y - S_y S_x)\beta = S_x(-\tfrac{1}{2}i\alpha) - S_y(\tfrac{1}{2}\alpha) \;=\; -\tfrac{1}{2}i(\tfrac{1}{2}\beta) - \tfrac{1}{2}(\tfrac{1}{2}i\beta) = -\tfrac{1}{2}i\beta = iS_z\beta.$$

In other words, $(S_x S_y - S_y S_x)$ and iS_z have exactly the same effect on each basis vector and hence on *any* vector in spin space; the two operators are *equal*. Cyclic permutation of subscripts, corresponding to use of the other anticommutators in (4.37), gives the full set of commutation relations:

COROLLARY 13. The spin operators for an electron satisfy the commutation relations

$$S_x S_y - S_y S_x = iS_z,$$
$$S_y S_z - S_z S_y = iS_x, \tag{4.40}$$
$$S_z S_x - S_x S_z = iS_y,$$

which are a necessary consequence of the isotropy of space under rotations of the coordinate axes.

Had we made the opposite choice of signs in (4.39), we should have obtained the same relations with $-S_z$ in place of S_z; reversal of the z-component corresponds merely to the change from a right-handed to a left-handed coordinate system.

The commutation relations for spin are evidently quite different from those for the position and momentum variables. It may be shown, however, that they are formally identical with those for the operators associated with *orbital angular momentum*; they characterize the properties of angular momentum in general, a subject fully dealt with in Vol. 2 (Chapters 2 and 5).

Inclusion of spin in the wave function

Let us now consider how the description of spin can be absorbed into the wave function describing the *spatial* motion of a particle. We first use the Hamiltonian operator (1.21), which contains no spin terms. This will possess eigenvalues and eigenfunctions $\{E_i, \phi_i\}$ such that

$$\mathsf{H}\phi_i = E_i\phi_i.$$

In a state of maximal knowledge we require the state vector to be an eigenvector of a full set of commuting operators. Only *one* of the spin operators, arbitrarily taken as that for the z-axis, can be included in this set; and S_z must commute with H because S_z only works on linear combinations of α and β while H only works on linear combinations of $\{\phi_i\}$—and hence the order in which they operate is irrelevant.

The simultaneous eigenvectors of H and S_z are self-evident; if we take $\psi_i = \phi_i\alpha$ we obtain (cf. p. 96)

$$\mathsf{H}\psi_i = E_i\psi_i,$$

$$\mathsf{S}_z\psi_i = \tfrac{1}{2}\psi_i,$$

so ψ_i is a simultaneous eigenvector of the commuting operators for energy and spin z-component. Similarly $\bar{\psi}_i = \phi_i\beta$ is a simultaneous eigenvector but with opposite spin.

The set of all products $\{\phi_i\eta_j\}$, where $\eta_1 = \alpha$ and $\eta_2 = \beta$ is said to span a "product space" (Section 3.8) the complete sets involved being ∞-dimensional and two-dimensional respectively. Any state involving the space and spin description of a single spin $\tfrac{1}{2}$ particle may thus be expanded in the form

$$\psi = \sum_{ij}c_{ij}\phi_i\eta_j = \sum_i a_i\phi_i\alpha + \sum_i b_i\phi_i\beta = \phi_\alpha\alpha + \phi_\beta\beta,$$

the two sums corresponding to $j = 1, 2$ respectively (α and β terms). Thus

$$\psi = \phi_\alpha\alpha + \phi_\beta\beta. \tag{4.41}$$

This most general "two-component" wave function has characteristic transformation properties under rotation of the axes of spin quantization and is called a "spinor". Usually it is possible to work in terms of states of *definite* spin and then only one component is non-zero; the term *spin-orbital* is then usually applied to the single products $\phi\alpha$ or $\phi\beta$. Clearly Postulate 3 must now be interpreted to mean that, for a one-electron system, the state vector ψ may always be expanded in terms of a complete set of spin-orbitals; there is a similar interpretation for a many-electron system, the expansion of many-electron wave functions being fully considered elsewhere (Vol. 4).

It remains only to show that, by a simple notational device, the description of both spin and orbital motion may be treated in exactly the same way. At present, an orbital state ϕ has a concrete realization, in the Schrödinger language, as an element of function space $\phi(\mathbf{r})$, in which both the variable $\mathbf{r}$ and the corresponding value of $|\phi(\mathbf{r})|^2$ have a direct physical meaning; on the other hand, α and β are vectors in a two-dimensional space and there has been no need, so far, to introduce a "spin variable". The consequences of this difference are a little awkward. For example, in the state (4.41) the normalization integral means

$$\langle\psi|\psi\rangle = \langle(\phi_\alpha\alpha + \phi_\beta\beta)|\phi_\alpha\alpha + \phi_\beta\beta\rangle.$$

On expanding, and using the Schrödinger language, a typical term is

$$\langle\phi_\alpha\alpha|\phi_\beta\beta\rangle = [\int\phi_\alpha{}^*(\mathbf{r})\phi_\beta(\mathbf{r})d\mathbf{r}]\langle\alpha|\beta\rangle.$$

Since the spin states are described by orthonormal vectors, the result is

$$\langle\psi|\psi\rangle = \int\phi_\alpha{}^*(\mathbf{r})\phi_\alpha(\mathbf{r})d\mathbf{r} + \int\phi_\beta{}^*(\mathbf{r})\phi_\beta(\mathbf{r})d\mathbf{r}. \tag{4.42}$$

In other words, in a scalar product we must integrate over continuous variables and sum over different components.

There are various ways of avoiding this inconvenience. Here we shall formally introduce "spin functions" $\alpha(s)$, $\beta(s)$ and write the scalar products (4.33) as

$$\langle\alpha|\alpha\rangle = \int\alpha^*(s)\alpha(s)ds = 1, \quad \langle\alpha|\beta\rangle = \int\alpha^*(s)\beta(s)ds = 0, \text{ etc.} \tag{4.43}$$

We now adopt a very useful convention, using $\mathbf{x}$ to stand for both space and spin variables and reserving $\mathbf{r}$ for space and s for spin variables separately. The wave function, including spin, will then be

$$\psi(\mathbf{x}) = \psi(\mathbf{r}, s) = \phi_\alpha(\mathbf{r})\alpha(s) + \phi_\beta(\mathbf{r})\beta(s) \tag{4.44}$$

and all scalar products can then be indicated formally as integrals, without any summations over components:

$$\langle\psi|\psi\rangle = \int\psi^*(\mathbf{x})\psi(\mathbf{x})d\mathbf{x}. \tag{4.45}$$

The conventions ensure, of course, that on inserting (4.44) and performing spin integrations we shall retrieve the component form (4.42); but we may not wish to do this (e.g. in a complicated many-electron problem) until the end of a calculation.

Finally, we must confirm that the statistical interpretation of the wave function according to (4.8) continues to make sense when spin is included among the variables. First of all we note that the expectation value of any function $f(\mathbf{r})$ (of spatial variables) in the state ψ given in (4.41), becomes, on using the orthonormality of α and β,

$$\langle\psi|f(\mathbf{r})|\psi\rangle = \int f(\mathbf{r})[|\phi_\alpha(\mathbf{r})|^2 + |\phi_\beta(\mathbf{r})|^2]d\mathbf{r}. \tag{4.46}$$

This implies, as in (4.7) et seq., that the probability of finding the electron in *spatial* volume element $d\mathbf{r}$ is

$$[|\phi_\alpha(\mathbf{r})|^2 + |\phi_\beta(\mathbf{r})|^2]d\mathbf{r}.$$

But in any state $a\alpha + b\beta$ the relative probabilities of the two spin situations are in the ratio $|a|^2:|b|^2$. The two terms in the expression thus give the relative probabilities of the particle in $d\mathbf{r}$ having spin "up" or spin "down"; while their sum gives the probability of the particle being there with either spin.

It is possible to interpret the spin *functions* $\alpha(s)$ and $\beta(s)$, and the variables, in such a way that (4.44) leads to exactly the same conclusions.

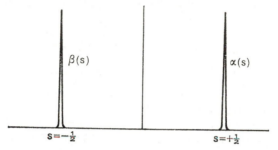

Fig. 4.1. Schematic representation of spin functions. $\alpha(s)$, $\beta(s)$ may be regarded formally as functions of a spin variable $s(=S_z)$, having the form of infinitely sharp "spikes" in the vicinity of $s = +\frac{1}{2}$ and $s = -\frac{1}{2}$ respectively.

To do this we might take s to be the value of the z-component of spin and suppose $|\alpha(s)|^2 ds$ indicates the probability of finding spin component in the range $(s, s + ds)$ just as $|\phi(\mathbf{r})|^2$ is that of finding position variables in the range $(\mathbf{r}, \mathbf{r} + d\mathbf{r})$. It is then evident that $\alpha(s)$ must vanish except when $s \simeq \frac{1}{2}$, subject to the normalization condition $\int|\alpha(s)|^2 ds = 1$. Such a function may be envisaged as the limit of a sharp "spike" in the vicinity of $s = \frac{1}{2}$, while $\beta(s)$ would be a similar "spike" vanishing except for $s \simeq -\frac{1}{2}$ (Fig. 4.1); there are mathematical difficulties in giving a literal interpretation to such "delta functions" (which are used again in Chapter 5) but from a formal point of view they are often very convenient. Here, for example, the statement (4.8) implies that (since $\alpha^*(s)\beta(s)$ cannot be non-zero)

$$|\psi(\mathbf{x})|^2 d\mathbf{x} = \{|\phi_\alpha(\mathbf{r})|^2|\alpha(s)|^2 + |\phi_\beta(\mathbf{r})|^2|\beta(s)|^2\}d\mathbf{r}\,ds = \begin{bmatrix} \text{Probability of} \\ \text{particle in } d\mathbf{r}, \\ \text{with spin in} \\ \text{the range} \\ (s, s + ds) \end{bmatrix}.$$

The probability of finding the particle in $d\mathbf{r}$, whatever the value of s, is obtained by integration ("summation" for all possible s values) and is thus

$$\{|\phi_\alpha(\mathbf{r})|^2 + |\phi_\beta(\mathbf{r})|^2\}d\mathbf{r}$$

exactly as above. But the first term, going with the $|\alpha(s)|^2$ factor comes entirely from the region $s \simeq +\frac{1}{2}$, while the second term comes from the region $s \simeq -\frac{1}{2}$. In other words, when the spin z component is used formally as a continuous variable the interpretation of the wave function according to (4.8) remains acceptable, provided the spin functions are defined so as to satisfy (4.43) and interpreted so that $\alpha(s) = 0$ $(s \neq \frac{1}{2})$, $\beta(s) = 0$ $(s \neq -\frac{1}{2})$. Although we have considered explicitly only a one-particle system the argument applies equally in the general case. Thus, for example, if a two-electron system is described by a wave function $\Psi(\mathbf{x}_1, \mathbf{x}_2)$, built up from a complete set of spin-orbitals, then

$$[\int|\Psi(\mathbf{x}_1, \mathbf{x}_2)|^2 ds_1 ds_2]d\mathbf{r}_1 d\mathbf{r}_2 = \begin{bmatrix} \text{Probability of particle 1 in } d\mathbf{r}_1, \\ \text{particle 2 in } d\mathbf{r}_2, \text{ without} \\ \text{reference to spin} \end{bmatrix}.$$

In other words, the probability density function for finding any spatial configuration of particles, irrespective of their particular spins, is obtained from $|\Psi|^2$ simply by a spin "integration". The need to pick out and sum over discrete spin components is thus eliminated and all variables can be handled in a formally similar way.

REFERENCES

COURANT, R. and HILBERT, D. (1953) *Methods of Mathematical Physics*, Vol. I, Interscience, New York.

DIRAC, P. A. M. (1947) *The Principles of Quantum Mechanics*, 3rd ed., Clarendon Press, Oxford.

KATO, T. (1951) *Trans. Am. Math. Soc.* **70**, 195.

KEMBLE, E. C. (1937) *The Fundamental Principles of Quantum Mechanics*, McGraw Hill, New York (reprinted 1958 by Dover, New York).

MESSIAH, A. (1961) *Quantum Mechanics*, Vols. I and II, North Holland, Amsterdam.

SCHIFF, L. (1968) *Quantum Mechanics*, McGraw Hill, New York.

TEMPLE, G. (1934) *Quantum Theory*, Methuen, London.

TOLMAN, R. L. (1938) *The Principles of Statistical Mechanics*, University Press, Oxford.

VAN NEUMANN, J. (1955) *Mathematical Foundations of Quantum Mechanics*, Princeton University Press, Princeton (translated from the German edition by R. T. Beyer).

CHAPTER 5

GENERAL THEORY
OF REPRESENTATIONS

5.1. Dirac notation. Discrete case

In Chapter 4 the postulates were formulated generally enough to provide a basis for the whole of non-relativistic quantum mechanics, but so far we are familiar with only one mathematical realization of the vectors and operators that occur throughout the theory. In Schrödinger language the vectors are elements of a Hilbert space comprising all (well-behaved) functions of particle coordinates, and the operators are partial differential operators working on these functions. Other possibilities clearly exist, however. The commutation relations (4.30) are equally well satisfied by the association

$$\mathsf{P}x \to p_x, \quad \mathsf{x} \to -\frac{\hbar}{i}\frac{\partial}{\partial p_x}, \text{ etc.,} \tag{5.1}$$

and this suggests a description in terms of functions of the *momentum* components, the operators associated with momentum becoming the multipliers, those associated with spatial coordinates becoming differential operators working on the momentum variables. We shall find presently that the Schrödinger and momentum languages are indeed equivalent and equally acceptable. First, however, we shall examine in a general way the possibility of passing from one type of description to another; this *transformation theory* is most easily formulated by starting from the matrix representation of the operator equations and using the Dirac notation (used so far only to indicate scalar products), which must now be explained in its full generality.

Let us introduce a discrete basis $\{\Phi_i\}$ such that

$$\Psi = \sum_i c_i \Phi_i. \tag{5.2}$$

The key equations referring to (i) the action of an operator on a vector, and (ii) the product of two operators, may then be transcribed into a matrix form fully discussed in Section 3.5. The basic equations become

$$\Psi' = A\Psi, \quad c' = Ac, \quad c_i' = \sum_j A_{ij}c_j, \tag{5.3a}$$

$$C = AB, \quad C = AB, \quad C_{ij} = \sum_k A_{ik}B_{kj}, \tag{5.3b}$$

where the three statements on each line are entirely equivalent. In the matrix form (second statement) c and c' are sets of expansion coefficients representing Ψ and Ψ' (collected into column matrices), while A, B and C are square matrices representing the operators A, B and C. The matrix equations may be written in subscript form (third statement) where A_{ij}, for example, is the element in the ith row and jth column of matrix A. We shall ignore all points of mathematical rigour, assuming that the basis is complete and that infinite matrices can be handled without regard to questions of convergence, etc.

The individual matrix elements c_i, A_{ij}, etc., are expressible as scalar products, and provided the basis is orthonormal (a condition which must be imposed in nearly all that follows) we obtain as in Section 3.5, equations (3.48) and (3.47),

$$c_i = \langle \Phi_i | \Psi \rangle, \quad A_{ij} = \langle \Phi_i | A | \Phi_j \rangle, \tag{5.4}$$

where A_{ij} is essentially the scalar product $\langle \Phi_i | A\Phi_j \rangle$ and the second vertical stroke is inserted (cf. p. 63) merely for notational convenience. So far, the pointed-bracket notation has been used only to indicate the Hermitian scalar product. In Dirac's use of the notation, however, which we now adopt more fully, each part of the scalar products yielding c_i and A_{ij} is regarded as a distinct entity: in general

$|\Psi\rangle$ is a vector in Hilbert space, indicated previously by Ψ;

$|\Phi_i\rangle$ is a basis vector in Hilbert space, indicated previously by Φ_i;

$\langle \Phi_i |$ is called the "dual vector" associated with $|\Phi_i\rangle$, indicated by $\Phi_i{}^*$ in the symbolic notation used in (3.44);

$A|\Psi\rangle$ is the vector arising when operator A is applied to $|\Psi\rangle$, indicated previously by $A\Psi$;

$\langle \Phi_i | \Phi_j \rangle$ is the scalar product of Φ_i and Φ_j, written alternatively as $\Phi_i{}^*\Phi_j$ when the Hermitian scalar product was first introduced in (3.44).

The symbols $|\,\rangle$ and $\langle\,|$ are referred to as "ket" and "bra", respectively, and a specific ket or bra is indicated by writing its name inside, e.g. $|\Phi_i\rangle$ or simply $|i\rangle$. When ket and bra come together in the order $\langle\,|\,\rangle$, to form a bra(c)ket, the scalar product is implied. Thus $A_{ij} = \langle \Phi_i | A | \Phi_j \rangle$ conveniently indicates the sequence of operations by which the matrix element may be obtained: reading from right to left, we take the ket

$|\Phi_j\rangle$, operate with A to get a new ket, and complete the bra(c)ket with $\langle\Phi_i|$. This is a particularly useful notation when the vectors require several labels (e.g. quantum numbers specifying the various eigenstates).

With Dirac notation the operator and component forms of (5.3a) and (5.3b) become, using (5.4),

$$|\Psi'\rangle = A|\Psi\rangle, \qquad \langle\Phi_i|\Psi'\rangle = \sum_j \langle\Phi_i|A|\Phi_j\rangle\langle\Phi_j|\Psi\rangle \qquad (5.5a)$$

$$C = AB, \qquad \langle\Phi_i|C|\Phi_j\rangle = \sum_k \langle\Phi_i|A|\Phi_k\rangle\langle\Phi_k|B|\Phi_j\rangle. \qquad (5.5b)$$

The forms on the right clearly follow from those on the left simply by inserting basis vector bra's and ket's on both sides of an equation, to complete the brackets defining components and matrix elements; in labelling the bra's and ket's those which come together in the order ket–bra (e.g. the $|\Phi_k\rangle\langle\Phi_k|$ in (5.5b) must bear the same label and a corresponding summation is implied (e.g. Σ_k in (5.5b)). In this way the "chain-rule" for the subscripts in matrix multiplication is automatic; for example, abbreviating $|\Phi_i\rangle$ to $|i\rangle$, etc.,

$$[A\ B\ C]_{il} = \sum_j\sum_k A_{ij}B_{jk}C_{kl} = \sum_{j,k} \langle i|A|j\rangle\langle j|B|k\rangle\langle k|C|l\rangle.$$

It should also be noted that even a ket–bra product such as $|\Phi_k\rangle\langle\Phi_k|$ has an interpretation of its own; it is the Dirac notation for the "dyad" $\Phi_k\Phi_k{}^*$ introduced in Appendix 4 and interpreted as a *projection operator*. Thus,

$$\Psi' = (\Phi_k\Phi_k{}^*)\Psi = \Phi_k(\Phi_k{}^*\Psi) = \Phi_k\langle\Phi_k|\Psi\rangle,$$

shows that the operator turns Ψ into a multiple of Φ_k, and this becomes in Dirac notation

$$|\Psi'\rangle = |\Phi_k\rangle\langle\Phi_k|\Psi\rangle,$$

the meaning being the same whichever way the three symbols are paired. The "resolution of the identity" (Appendix 4, p. 148) in terms of projection operators shows that

$$\sum_k |\Phi_k\rangle\langle\Phi_k| = I \qquad (5.6)$$

and gives another interpretation of the right-hand side of (5.5b); the unit operator may be inserted between two operators without changing the expression and hence

$$\langle\Phi_i|C|\Phi_j\rangle = \langle\Phi_i|AB|\Phi_j\rangle = \sum_k \langle\Phi_i|A|\Phi_k\rangle\langle\Phi_k|B|\Phi_j\rangle.$$

It is evident that the Dirac notation provides an almost "fool-proof" method of writing operator equations in the matrix language appropriate to any chosen basis. We now consider an example, partly to illustrate the use of the notation and partly to show how the basic equations of quantum mechanics (Chapter 4) may be transcribed into matrix language and used in obtaining a direct algebraic solution of certain simple problems, quite independently of Schrödinger's equations.

5.2. An example. The harmonic oscillator

Let us consider the oscillator with Hamiltonian operator (Section 2.2)

$$H = \frac{p^2}{2m} + \tfrac{1}{2}m\omega^2x^2,$$

where $\omega/2\pi$ would be the classically expected natural frequency of oscillation. If we introduce a discrete basis, provided by the eigenvectors of H, operator equations are replaced by formally identical matrix equations—yielding the "matrix mechanics" approach of Heisenberg, Born and Jordan.

The eigenvalue equation $H\Psi = E\Psi$ becomes, in Dirac notation,

$$H|\rangle = E|\rangle$$

with energy eigenvectors $|E_0\rangle$, $|E_1\rangle$, ... which satisfy $H|E_j\rangle = E_j|E_j\rangle$. The matrix H associated with H has elements $\langle E_i|H|E_j\rangle = E_j\langle E_i|E_j\rangle = E_i\delta_{ij}$ (since the eigenvectors may be assumed orthonormal (p. 88)) and is therefore diagonal.

It is convenient to express H in terms of the new operators

$$A = p - im\omega x \quad A^\dagger = p + im\omega x$$

noting that $A^\dagger$ is the Hermitian adjoint of A, by (3.33). For then

$$A^\dagger A = p^2 + im\omega(xp - px) + m^2\omega^2x^2$$

and from the commutation relation (4.30), $xp - px = i\hbar I$, this gives

$$H = \frac{1}{2m}(A^\dagger A + \hbar m\omega I).$$

The properties of the Hamiltonian are characterized by the commutation rules for the three operators H, x, p; these may be derived purely algebraically, starting from that for x and p. In this way we find, using the linear combinations defining A and $A^\dagger$,

$$HA - AH = \hbar\omega A, \quad HA^\dagger - A^\dagger H = \hbar\omega A^\dagger.$$

Let us now write the first commutator equation in matrix form, using

Dirac notation for the elements. Since H is diagonal, with $\langle E_i|H|E_j\rangle = E_i\delta_{ij}$, the ij element of the first equation is

$$E_i\langle E_i|A|E_j\rangle - \langle E_i|A|E_j\rangle E_j = \hbar\omega\langle E_i|A|E_j\rangle$$

and therefore

$$(E_i - E_j - \hbar\omega)\langle E_i|A|E_j\rangle = 0.$$

This shows that either (i) $E_i = E_j + \hbar\omega$, or (ii) $\langle E_i|A|E_j\rangle = 0$. The matrix associated with A thus has non-zero elements only between states whose energies differ by an amount $\hbar\omega$. We label these states in ascending numerical order so that

$$E_1 = E_0 + \hbar\omega, \quad E_2 = E_0 + 2\hbar\omega, \ldots \quad E_n = E_0 + n\hbar\omega, \ldots$$

That there is a *least* eigenvalue E_0 follows from the form of the Hamiltonian, which gives

$$E_i = \langle E_i|H|E_i\rangle = (1/2m)[\langle E_i|A^\dagger|E_j\rangle\langle E_j|A|E_i\rangle + \hbar m\omega]$$

and is therefore essentially positive unless $|E_i\rangle$ vanishes.‡

To determine the value of E_0 we put $i = 0$ and find, since $E_j > E_0$ and hence $(E_i - E_j - h\omega) \neq 0$, that all matrix elements $\langle E_0|A|E_j\rangle$ must vanish. The above expression for E_i then gives

$$E_0 = (1/2m)\hbar m\omega$$

and the complete sequence of eigenvalues thus becomes

$$E_n = (n + \tfrac{1}{2})h\nu \quad (n = 0, 1, 2, \ldots)$$

where $\nu = \omega/2\pi$ is the classically expected frequency of oscillation. This result is in complete agreement with that obtained by using Schrödinger's representation (Section 2.2) and solving Hermite's differential equation, but has been obtained by purely algebraic arguments, directly from the general postulates of Chapter 4.

The solution is completed by obtaining the matrices associated with x and p. Those associated with A and $A^\dagger$ contain only one element in each row and we find $|\langle E_n|A^\dagger|E_{n-1}\rangle|^2 = 2nmh\nu$; apart from a unimodular phase factor (which is arbitrary and may be dropped). This makes $\langle E_n|A^\dagger|E_{n-1}\rangle = \sqrt{(2mh\nu)}\sqrt{n}$ and it is then inferred that the matrices associated with x and p are

---

‡$A^\dagger$ is the adjoint of A and hence (see p. 68) $\langle\Psi|A^\dagger A\Psi\rangle = \langle A\Psi|A\Psi\rangle > 0$ ($\Psi \neq 0$). The first term in square brackets is $\langle E_i|A^\dagger A|E_i\rangle$ (only one term non-zero in the matrix product).

$$\mathbf{x} = \frac{1}{2\pi}\sqrt{\left(\frac{h}{2mv}\right)}\begin{bmatrix} 0 & \sqrt{1} & 0 & 0 & .. \\ -\sqrt{1} & 0 & \sqrt{2} & 0 & .. \\ 0 & -\sqrt{2} & 0 & \sqrt{3} & .. \\ 0 & 0 & -\sqrt{3} & 0 & .. \\ .... \end{bmatrix}$$

$$\mathbf{p} = \sqrt{\left(\frac{mhv}{2}\right)}\begin{bmatrix} 0 & \sqrt{1} & 0 & 0 & .. \\ \sqrt{1} & 0 & \sqrt{2} & 0 & .. \\ 0 & \sqrt{2} & 0 & \sqrt{3} & .. \\ 0 & 0 & \sqrt{2} & 0 & .. \\ .... \end{bmatrix}$$

It is easily confirmed that these matrices satisfy the commutation rule $\mathbf{xp} - \mathbf{px} = i\hbar\mathbf{I}$ where $\mathbf{I}$ is the unit matrix.

5.3.   Dirac notation. Continuous case

When matrices are associated with operators, in the way just described, we obtain a *representation* in the usual sense of vector space theory; but the term is very frequently used in quantum mechanics in a broader sense, including the case where the basis vectors are not discrete but are labelled by continuous variables. The equations of the last section must then be rewritten and it becomes more natural to use the language of the theory of integral equations. We shall not develop the theory of "continuous representations" in much detail; it is sufficient for present purposes to indicate in a formal way the necessary transcription of the matrix equations, and to obtain the general rules for passing between one quantum mechanical language and another.

In the harmonic oscillator example we took a set of energy eigenvectors as a discrete basis. Suppose, however, we had taken the set of momentum eigenvectors for a free particle (still, for simplicity, in one dimension). We know from Section 2.4 that momentum eigenstates exist for all p values in the interval $(-\infty, +\infty)$; but that we can make the spectrum of eigenvalues of p discrete by the artifice of putting the system in a large box with periodic boundary conditions (p. 36), thus bringing the states into one-to-one correspondence with the integers $(-\infty < n < +\infty)$. In this case we can examine how the discrete representation passes over into a continuous one as the box is made infinitely large. Starting from the eigenvectors $|p_n\rangle$ we write the eigenvalue equation $\mathsf{H}|\rangle = E|\rangle$, whose solutions may be denoted by $|\rangle = |E_1\rangle, |E_2\rangle, \dots$ (labelled by the energy eigenvalues), in the form

$$\sum_j \langle p_i|\mathsf{H}|p_j\rangle\langle p_j|\rangle = E\langle p_i|\rangle \tag{5.7}$$

where $\langle p_i| \rangle$ is the component of $| \rangle$ along the basis vector $|p_i\rangle$. If $| \rangle$ is a normalized vector this means

$$\sum_i |c_i|^2 = \sum_i \langle |p_i\rangle\langle p_i| \rangle = 1 \qquad (5.8)$$

where $|c_i|^2$ is the probability in state $| \rangle$ that measurement of momentum will yield the value p_i.

To pass to the limit we simply note that the summand in (5.8) becomes a continuous function of p. The term for $p = p_n$ may be written $f(p_n)^*f(p_n)\Delta$ where $f(p_n)^*f(p_n)$ is a *density function* (the contribution per unit increase in momentum) evaluated at $p = p_n$, and $\Delta (= \hbar/L)$ is the momentum increase in going from p_n to p_{n+1}. If we momentarily introduce the notation $f(p) = (p|)$, $f^*(p) = (|p)$ and pass to the limit $L \to \infty$ (infinitely large box), so that Δ becomes the differential dp, the sum (5.8) goes over into an integral:

$$\sum_i \langle |p_i\rangle\langle p_i| \rangle = \sum_i (|p_i)\Delta(p_i|) \to \int (|p)dp(p|).$$

In the case of the momentum variables this limiting procedure may be rigorously justified by Fourier transform theory; more generally, however, we assume that sums may be freely translated into integrals unless we find evidence to the contrary. We also revert to the pointed bracket notation, using $\langle p| \rangle$ instead of $(p|)$, and note that this amounts to a *redefinition* of $\langle p| \rangle$ when p is a continuous variable: $|\langle p| \rangle|^2$ is now a probability *per unit momentum range* of measurement yielding a value p.

All the equations relating to discrete representations may be transcribed similarly. For example, equation (5.7) takes the form

$$\int \langle p|\mathsf{H}|p'\rangle\langle p'| \rangle dp' = E\langle p| \rangle. \qquad (5.9)$$

Thus the components of the vector $| \rangle$ are replaced by a *continuous* function of p, where p plays the part of a basis vector index, and the matrix elements of H are replaced by a function of *two* continuous variables p, p'. The function $\langle p|\mathsf{H}|p'\rangle$ is sometimes regarded as the element of a "continuous matrix" since p and p' play the part of row and column indices: this analogy is useful, but more properly $\langle p|\mathsf{H}|p'\rangle$ is an *integral kernel* and (5.9) is an integral equation. In general a kernel $H(x;x')$ describes an operator H according to the convention

$$f'(x) = \mathsf{H}f(x) = \int H(x;x')f(x')dx', \qquad (5.10)$$

i.e. $f'(x)$ is obtained from $f(x)$ by changing the variable from x to x', multiplying by the kernel, and integrating over x' to obtain a new function of x, namely $f'(x)$. In the present instance, the state vector $| \rangle$ is described by means of a function of momentum; $\langle p| \rangle$ is referred to as the wave function in "momentum space" or in the "momentum representation"; and (5.9) is the eigenvalue equation in the "momentum representation".

The extension to several variables is formally straightforward; in three dimensions we may choose the *three* compatible variables, p_x, p_y, p_z and associate with a state vector $|\rangle$ its momentum function $\langle p_x p_y p_z|\rangle$; if there are many particles we should use the set of *all* momentum components writing the momentum function, for brevity, as $\langle \mathbf{p}_1 \mathbf{p}_2 \ldots \mathbf{p}_N|\rangle$ or even $\langle \mathbf{p}|\rangle$. When a vector symbol is used to denote a set of variables the corresponding $\int d\mathbf{p}$ means, of course, integration over all variables. It should be noted that the variables chosen must always form a compatible set (i.e. their operators must commute) since $|p_x p_y p_z\rangle$, for example, refers to a physically observable eigenstate of all three operators.

Since we already have considerable knowledge of one particular continuous representation, that employed in Schrödinger's theory, it is useful to consider at this point the relationship between the original formulation—using differential operators—and that of the present section, which appears to lead to integral equations. Again we consider first a one-dimensional system, a particle moving along the x-axis. An energy eigenvector $|\rangle$ is then characterized by the continuous function $\langle x|\rangle$ such that $|\langle x|\rangle|^2$ is the probability per unit range of finding the particle at x. In fact, $\langle x|\rangle$ is simply the Schrödinger wave function for state $|\rangle$. In the Schrödinger or "coordinate" representation the eigenvalue equation would evidently appear as an integral equation analogous to (4.7), namely

$$\int \langle x|\mathsf{H}|x'\rangle\langle x'|\rangle dx' = E\langle x|\rangle. \tag{5.11a}$$

Why does this equation look so different from the familiar Schrödinger equation? Previously we have written, using Dirac notation for the Schrödinger wave function $\psi(x)$,

$$\mathsf{H}\langle x|\rangle = -\frac{\hbar^2}{2m}\frac{d^2\langle x|\rangle}{dx^2} + V(x)\langle x|\rangle = E\langle x|\rangle. \tag{5.11b}$$

The right-hand sides of (5.11a) and (5.11b) are identical but the left-hand sides appear to be quite different in form.

To understand this difference, we note that the form (5.11a) merely expresses the correspondence with a matrix formulation, and that the integration appears solely for this purpose. Since $\int |x'\rangle dx'\langle x'|$ is the limiting form, as we pass to a continuous representation, of the unit operator (cf. (5.6)) it is in fact superfluous and the left-hand side of (5.11a) really means $\langle x|\mathsf{H}|\rangle$, i.e. the function of x describing the components of $\mathsf{H}|\rangle$ in the x-representation, written in (5.11b) as

$H\langle x|\rangle$. The effect of the operators in H is well known, it is only by disguising them as integral operators that we make them look strange. Evidently to operate with the potential energy term in H we must strike out the integration over x' in (5.11a), giving x' the value x, and then simply multiply by the function $V(x)$; we must take

$$\int \langle x|V|x'\rangle dx'\langle x'| \rangle = V(x)\langle x|\rangle. \tag{5.12}$$

Similarly, to operate with the kinetic energy term T we must strike out the integration and then perform differentiations on $\langle x|\rangle$: we must take

$$\int \langle x|T|x'\rangle dx'\langle x'| \rangle = -\frac{\hbar^2}{2m}\frac{d^2}{dx^2}\langle x|\rangle. \tag{5.13}$$

In order to write the desired results in terms of integral kernels we introduce the Dirac delta function (already referred to in Section 4.9).

The delta function $\delta(x-x')$ is defined formally by the property

$$\int \delta(x-x')f(x')dx' = f(x), \tag{5.14}$$

where, as usual, integration is understood to be over the full range $(-\infty, +\infty)$: regarded as an integral kernel (cf. (5.10)) it must represent the *unit operator* that leaves any function $f(x)$ unchanged. This function, regarded by pure mathematicians as a monstrosity, must vanish when x' differs significantly from x, so as to eliminate contributions with $f(x')$ taking values other than $f(x)$. But it must be normalized so that

$$\int \delta(x-x')f(x')dx' \simeq f(x)\int \delta(x-x')dx' = f(x),$$

where $f(x')$ is assumed to vary slowly, compared with $\delta(x-x')$, in the region where the latter is non-zero. The delta function may thus be regarded as the limiting form of a function with a very sharp peak at $x' = x$ (cf. Fig. 4.1), the width becoming indefinitely small while the area under the curve keeps the value unity. Various realizations of $\delta(x-x')$ exist, and are satisfactory for many purposes, but in most cases the function is introduced simply for notational convenience. The use of the delta function can always be avoided, but only at the cost of delicate and sophisticated mathematical analysis; in applied mathematics its use may be regarded as a convenient shorthand for exhibiting in a transparent way the connections among matrix equations, integral equations and differential equations. For a detailed discussion of the δ-function the reader is referred elsewhere (e.g. Messiah, 1961, vol. 1, appendix A). Here we mention just one convenient representation of the function:

$$\delta(x-x') = \frac{1}{2\pi}\int_{-\infty}^{+\infty} \exp\,[ik(x-x')]dk. \tag{5.15}$$

The justification for the use of this form has been widely discussed (see, for example, Dirac, 1947, p. 95, or Messiah *loc. cit.*) and is implicit in the theory of Fourier transforms.

In terms of $\delta(x-x')$ the integral kernels $\langle x|\mathsf{V}|x'\rangle$ and $\langle x|\mathsf{T}|x'\rangle$ are easily defined; if we take

$$\langle x|\mathsf{V}|x'\rangle = V(x)\,\delta(x-x'), \quad \langle x|\mathsf{T}|x'\rangle = -\frac{\hbar^2}{2m}\frac{d^2}{dx^2}\,\delta(x-x') \quad (5.16)$$

it is clear that (5.12) and (5.13) are satisfied. For example,

$$\int\langle x|\mathsf{V}|x'\rangle dx'\langle x'| = \int V(x)\,\delta(x-x')\,dx'\langle x'| = V(x)\langle x|$$

from the δ-function property. It is also clear that the kernels associated with the basic position and momentum operators are

$$\langle x|\mathsf{x}|x'\rangle = x\,\delta(x-x'), \quad \langle x|\mathsf{p}|x'\rangle = \frac{\hbar}{i}\frac{d}{dx}\,\delta(x-x') \quad (5.17)$$

and it is easily verified, by considering

$$\int\langle x|\mathsf{p}|x''\rangle dx''\langle x''|\mathsf{p}|x'\rangle dx'\langle x'|$$

that $\langle x|\mathsf{T}|x'\rangle$ takes the form inferred above.

To extend these results to many dimensions, we note that a three-dimensional delta function may be defined in a formally similar way

$$\int\delta(\mathbf{r}-\mathbf{r}')d\mathbf{r}'f(\mathbf{r}') = f(\mathbf{r}) \quad (5.18)$$

where $\mathbf{r}$ comprises all three spatial coordinates, and that if there are many particles then

$$\langle \mathbf{r}_1, \mathbf{r}_2, \ldots |\mathsf{x}_i|\mathbf{r}_1, \mathbf{r}_2, \ldots\rangle = x_i\prod_j \delta(\mathbf{r}_j-\mathbf{r}_j'), \quad (5.19a)$$

$$\langle \mathbf{r}_1, \mathbf{r}_2, \ldots |\mathsf{p}_x(i)|\mathbf{r}_1, \mathbf{r}_2, \ldots\rangle = \frac{\hbar}{i}\frac{\partial}{\partial x_i}\prod_j \delta(\mathbf{r}_j-\mathbf{r}_j'). \quad (5.19b)$$

Each factor in $\prod_j \delta(\mathbf{r}_j-\mathbf{r}_j')$ represents a unit operator for any function of $\mathbf{r}_j$, and the presence of the factors with $j \neq i$ simply means that the operators x_i and $\mathsf{p}_x(i)$ work only on the variables of particle i, in which case their forms are determined by (5.17). There is an integral representation of the three-dimensional $\delta(\mathbf{r}-\mathbf{r}')$ exactly analogous to (5.15), namely

$$\delta(\mathbf{r}-\mathbf{r}') = \frac{1}{(2\pi)^3}\int \exp[i\mathbf{k}\cdot(\mathbf{r}-\mathbf{r}')]d\mathbf{k}, \quad (5.20)$$

where $\mathbf{k}$ and $\mathbf{r}$ are vectors and the integration is over all space; this clearly amounts simply to a product of three factors of the type (5.15).

5.4. Transformation theory. The momentum representation

The different languages available for describing a quantum mechanical system correspond merely to different choices of basis in the Hilbert space of eigenfunctions. In the harmonic oscillator example (p. 107) we were able to work directly in the representation provided by energy eigenvectors $\{|E_i\rangle\}$, the corresponding operator H being represented by a diagonal matrix with element

$$\langle E_i|\mathsf{H}|E_j\rangle = E_i\,\delta_{ij}$$

while the other operators x and p, not commuting with H, had non-diagonal matrices.

In the Schrödinger representation we choose the "continuous basis" with eigenvectors $|x\rangle$: with a corresponding operator x we again associate a diagonal "matrix" with elements depending on the continuous variables x, x'

$$\langle x|\mathsf{x}|x'\rangle = x\,\delta(x-x').$$

Operators not commuting with x will have non-diagonal matrices.‡

On the other hand, we might equally well set up a "momentum representation" by choosing momentum eigenvectors $|p\rangle$ as a basis. The momentum operator will then have associated with it a diagonal "matrix" with elements

$$\langle p|\mathsf{p}|p'\rangle = p\,\delta(p-p')$$

and again other (non-commuting) operators will be represented by non-diagonal "matrices". *Transformation theory* is concerned with the relationship between representations arising from alternative choices of basis.

We start from the equations describing a basis change (or "rotation") in the case of finite vector spaces (see Section 3.6). If rotation is described by an operator U the bases are related by $\overline{\Phi}_i = \mathsf{U}\Phi_i$ and, provided the rotation is restricted so as to preserve orthonormality of the basis, this operator is of special form; it is *unitary* in the sense that its adjoint $\mathsf{U}^\dagger$ is at the same time its inverse. This follows from the condition that the scalar product of any two vectors shall be left unchanged when they suffer the same rotation. Thus the condition

---

‡There is a difficulty here in that, according to (5.17), the operator associated with p is also apparently "diagonal". A more careful examination of the limiting process in passing from a discrete to continuous spectrum shows that p in fact connects different states whose eigenvalues, in the limit, become infinitely close (cf. the quasi-diagonal form of the p matrix in the harmonic oscillator example, p. 109).

$$\langle \overline{\Phi}_i | \overline{\Phi}_j \rangle = \langle U\Phi_i | U\Phi_j \rangle = \langle \Phi_i | U^\dagger U\Phi_j \rangle = \langle \Phi_i | \Phi_j \rangle \quad (\text{all } i, j)$$

shows that

$$U^\dagger U = I.$$

The use of the term "unitary" is consistent with the fact (p. 67) that the matrix U representing such a rotation must itself be a *unitary matrix*; and it follows readily that $U^\dagger$ is represented by $U^\dagger$.

Let us now rewrite the transformation equations of Section 3.6, relating vector components and matrices relative to two different bases, in Dirac notation. Basis vectors, components of an arbitrary vector Ψ, and matrices transform according to

(i) *Basis vectors*

$$\Phi_i \to \overline{\Phi}_i = U\Phi_i = \sum_j \Phi_j U_{ji}.$$

(ii) *Components of arbitrary vector* Ψ:

$$c_i \to \bar{c}_i = \sum_j U^\dagger_{ij} c_j.$$

(iii) *Matrix elements of operator* A:

$$A_{ij} \to \overline{A}_{ij} = (U^\dagger A U)_{ij} = \sum_{k,l} U^\dagger_{ik} A_{kl} U_{lj}.$$

The first relation becomes, in Dirac notation,

(i)a $\quad |\Phi_i\rangle \to |\overline{\Phi}_i\rangle = U|\Phi_i\rangle = \sum_j |\Phi_j\rangle\langle\Phi_j|U|\Phi_i\rangle = \sum_j |\Phi_j\rangle\langle\Phi_j|\overline{\Phi}_i\rangle$

where the last form shows how the actual matrix elements of U may be eliminated by introducing the scalar products of vectors from the *different bases*. The same device may be used to eliminate matrix elements of $U^\dagger$. Let us now use an open ket to denote the arbitrary vector Ψ. The second relation then (ii) becomes

(ii)a $\quad \langle\Phi_i| \to \langle\overline{\Phi}_i| = \sum_j \langle\Phi_i|U^\dagger|\Phi_j\rangle\langle\Phi_j| = \sum_j \langle\overline{\Phi}_i|\Phi_j\rangle\langle\Phi_j|$

because $\quad \langle\overline{\Phi}_i|\Phi_j\rangle = \langle U\Phi_i|\Phi_j\rangle = \langle\Phi_i|U^\dagger|\Phi_j\rangle.$

Finally, the third relation takes the form

(iii)a $\quad \langle\Phi_i|A|\Phi_j\rangle \to \langle\overline{\Phi}_i|A|\overline{\Phi}_j\rangle = \sum_{k,l} \langle\Phi_i|U^\dagger|\Phi_k\rangle\langle\Phi_k|A|\Phi_l\rangle\langle\Phi_l|U|\Phi_j\rangle$

$$= \sum_{k,l} \langle\overline{\Phi}_i|\Phi_k\rangle\langle\Phi_k|A|\Phi_l\rangle\langle\Phi_l|\overline{\Phi}_j\rangle$$

where again the matrix elements of U are involved only implicitly in the final form. The beauty of the Dirac notation is that the final forms in (i)a–(iii)a could all have been written down without any effort by inserting unit operators of the form $\sum_k |\Phi_k\rangle\langle\Phi_k|$; in (iii)a, for instance, the final result follows on inserting unit operators on either side of the A in the required matrix element $\langle\overline{\Phi}_i|A|\overline{\Phi}_j\rangle$. The notation itself does the derivation.

The generalization to the continuous spectrum is formally straight-forward. If the kets $|\Phi_i\rangle$ are replaced by $|x\rangle$, the latter kets being labelled by a continuous index x (e.g. a position coordinate), and the $|\overline{\Phi}_i\rangle$ are replaced by kets $|p\rangle$, again labelled by a continuous index p (e.g. a momentum component), then the transformation equations (ii)a and (iii)a become

Wave function

$$\langle x|\rangle \to \langle p|\rangle = \int \langle p|x\rangle \, dx \langle x|\rangle. \tag{5.21}$$

Operator

$$\langle x|A|x'\rangle \to \langle p|A|p'\rangle = \int \langle p|x\rangle \, dx \langle x|A|x'\rangle \, dx' \langle x'|p'\rangle. \tag{5.22}$$

These are the basic laws of transformation theory. Their meaning should be considered carefully; for example, $\langle x|\rangle$ is the Schrödinger wave function for state $|\rangle$, while $\langle p|\rangle$ is a corresponding "momentum space" wave function; also $\langle x|p\rangle$ is the Schrödinger wave function for a state $|p\rangle$ of definite momentum, and this is the function which provides the bridge between the two languages. If the kets of each representation depend on several variables nothing is changed except that the integrations over single variables become integrations over several variables. To learn how to use the transformation theory we study a typical example.

5.5.   The Schrödinger equation in momentum space

In three dimensions the Hamiltonian contains kinetic and potential energy operators whose Schrödinger operators (in Dirac form corresponding to (5.16)) are

$$\langle \mathbf{r}|T|\mathbf{r}'\rangle = -\frac{\hbar^2}{2m}\left(\frac{\partial^2}{\partial x^2}+\frac{\partial^2}{\partial y^2}+\frac{\partial^2}{\partial z^2}\right)\delta(\mathbf{r}-\mathbf{r}') \tag{5.23}$$

$$\langle \mathbf{r}|V|\mathbf{r}'\rangle = V(\mathbf{r})\delta(\mathbf{r}-\mathbf{r}') \tag{5.24}$$

where, as usual, $\mathbf{r}$ stands for the triplet of position variables, and where $\delta(\mathbf{r}-\mathbf{r}')$ is the three-dimensional delta function of (5.18). The states of definite momentum, for one electron, have been found in Section 2.4 (see also the Example on p. 81) and are, using the normalization of (2.28),

$$\langle \mathbf{r}|p\rangle = (2\pi\hbar)^{-3/2}\exp(i\mathbf{p}\cdot\mathbf{r}/\hbar). \tag{5.25}$$

From the definition of the momentum representation $\langle \mathbf{p}|\mathsf{T}|\mathbf{p}'\rangle$ should be a diagonal matrix since it is a function of $\mathbf{p}$; but to check that the transformation equations work we may derive it from the Schrödinger form. Thus (remembering that, from (3.46), $\langle \mathbf{p}|\mathbf{r}\rangle = \langle \mathbf{r}|\mathbf{p}\rangle^*$),

$$\langle \mathbf{p}|\mathsf{T}|\mathbf{p}\rangle = \int \langle \mathbf{p}|\mathbf{r}\rangle d\mathbf{r} \left[-\frac{\hbar^2}{2m} \left(\frac{\partial^2}{\partial x^2} + \frac{\partial^2}{\partial y^2} + \frac{\partial^2}{\partial x^2} \right) \delta(\mathbf{r}-\mathbf{r}') \right] d\mathbf{r}' \langle \mathbf{r}'|\mathbf{p}'\rangle$$

$$= -\frac{\hbar^2}{2m} \frac{1}{(2\pi\hbar)^3} \int \exp\left(\frac{i}{\hbar} \mathbf{p} \cdot \mathbf{r} \right) \left(\frac{\partial^2}{\partial x^2} + \frac{\partial^2}{\partial y^2} + \frac{\partial^2}{\partial x^2} \right)$$

$$\times \exp\left[-\frac{i}{\hbar} (x p_x' + y p_y' + z p_z') \right] dx\, dy\, dz$$

$$= -\frac{\hbar^2}{2m} \frac{1}{(2\pi\hbar)^3} \exp\left(\frac{i}{\hbar} \mathbf{p} \cdot \mathbf{r} \right) \left(-\frac{1}{\hbar^2} (p_x'^2 + p_y'^2 + p_z'^2) \right)$$

$$\times \exp\left[-\frac{i}{\hbar} \mathbf{p}' \cdot \mathbf{r} \right] d\mathbf{r}$$

which may be written

$$\langle \mathbf{p}|\mathsf{T}|\mathbf{p}'\rangle = \frac{\mathbf{p}'^2}{2m} \frac{1}{(2\pi\hbar)^3} \int \exp\left[\frac{i}{\hbar} (\mathbf{p}-\mathbf{p}') \cdot \mathbf{r} \right] d\mathbf{r}.$$

The integral that remains, however, is simply a three-dimensional delta function $\delta(\mathbf{p}-\mathbf{p}')$ in the integral representation given in (5.20), as follows readily on changing the integration variable from $\mathbf{k}$ to $\mathbf{r}/\hbar$.‡ Hence, as was anticipated

$$\langle p|\mathsf{T}|p'\rangle = \frac{\mathbf{p}^2}{2m} \delta(\mathbf{p}-\mathbf{p}') \tag{5.26}$$

—the kinetic energy operator is thus equivalent to multiplication by $(\mathbf{p}^2/2m)$. The special value of the normalization adopted in (2.28) is now clear; it is often referred to as "delta-function normalization".

---

‡It is interesting to note that this integral may be written (from (5.25)) in the form

$$\int \langle \mathbf{p}|\mathbf{r}\rangle d\mathbf{r}\langle \mathbf{r}|\mathbf{p}'\rangle = \langle \mathbf{p}|\mathsf{I}|\mathbf{p}'\rangle$$

where I is a unit operator, defined in (5.6). Clearly

$$\int \langle \mathbf{p}|\mathsf{I}|\mathbf{p}'\rangle f(\mathbf{p}') d\mathbf{p}' = f(\mathbf{p}),$$

are required of the delta function.

It is the potential energy, in momentum language, that leads to the non-trivial operator: it becomes the integral operator whose kernel is (cf. (5.22))

$$\langle \mathbf{p}|\mathsf{V}|\mathbf{p}'\rangle = \int \langle \mathbf{p}|\mathbf{r}\rangle d\mathbf{r}\, V(\mathbf{r})\, \delta(\mathbf{r}-\mathbf{r}')d\mathbf{r}'\langle \mathbf{r}'|\mathbf{p}'\rangle$$

$$= \frac{1}{(2\pi\hbar)^3} \int V(\mathbf{r}) \exp\left[\frac{i}{\hbar}(\mathbf{p}-\mathbf{p}')\cdot\mathbf{r}\right]d\mathbf{r}. \qquad (5.27)$$

The potential energy kernel, in momentum space, is thus a Fourier transform of the potential energy function in ordinary space.

Finally we write the Schrödinger equation in momentum language; on denoting $\langle \mathbf{p}|\rangle$ by $\chi(\mathbf{p})$ it reads

$$(\mathbf{p}^2 - \mathbf{p}_0{}^2)\chi(\mathbf{p}) = -2m \int \langle \mathbf{p}|\mathsf{V}|\mathbf{p}'\rangle\chi(\mathbf{p}')d\mathbf{p}' \quad (\mathbf{p}_0{}^2 = 2mE). \qquad (5.28)$$

Such equations can sometimes be solved using methods from the theory of integral equations but to review such developments would take us too far afield.

The transformation we have employed is easily generalized to many particles by interpreting $\mathbf{r}$ and $\mathbf{p}$ as the sets $\mathbf{r}_1, \mathbf{r}_2, \ldots \mathbf{r}_N$ and $\mathbf{p}_1, \mathbf{p}_2, \ldots \mathbf{p}_N$, respectively, and using $\langle \mathbf{r}_1, \mathbf{r}_2, \ldots \mathbf{r}_N|\mathbf{p}_1, \mathbf{p}_2, \ldots \mathbf{p}_N\rangle = \langle \mathbf{r}_1|\mathbf{p}_1\rangle \ldots \langle \mathbf{r}_N|\mathbf{p}_N\rangle$. However, there appear to be few special advantages of working in momentum space, and when momentum wave functions are required (as for example in the interpretation of Compton profiles in scattering experiments) it is more usual to obtain them indirectly by using (5.21), or its many-variable counterpart, to transform a given Schrödinger wave function. This is the procedure first employed by Pauling and Podolsky (1929), for hydrogen-like wave functions, used for molecular functions by Coulson and Duncanson (1941) and recently by Henneker and Cade (1968), Epstein (1970).

5.6. Time-evolution. The Heisenberg representation

Up to this point, we have always considered that the operators associated with physical quantities that do not involve time explicitly are themselves time-independent, and that only the state vector depends on time. This interpretation is implicit in the time-dependent Schrödinger equation (noting that the abstract ket depends only on t)

$$-\frac{\hbar}{i}\frac{d}{dt}|\Psi\rangle = \mathsf{H}|\Psi\rangle \qquad (5.29)$$

and is therefore characteristic of Schrödinger's formulation of quantum mechanics, even though we may use kets or vectors so as not to be committed to any particular language. When we consider the time-evolution of a system more generally, however, it turns out that other interpretations are open to us; one of these yields the "Heisenberg representation".

First we consider in general the time-development arising from (5.29). If the state at time $t = t_0$ is denoted by $\left|\Psi(t_0)\right\rangle$ we shall write

$$\left|\Psi(t)\right\rangle = \mathsf{U}(t, t_0)\left|\Psi(t_0)\right\rangle \tag{5.30}$$

and call $\mathsf{U}(t, t_0)$ the "evolution operator". Since $\left|\Psi(t+\delta t)\right\rangle$ is obtained from $\left|\Psi(t)\right\rangle$ by means of a linear operator (H) we infer U is also linear; from the normalization condition $\left\langle \mathsf{U}\Psi|\mathsf{U}\Psi\right\rangle = 1$ it may also be concluded (p. 114) that U is a unitary operator. In the Schrödinger picture, the states of a system are thus described by unit vectors which change as time passes. In particular, when H does not contain the time, there are certain stationary eigenstates of energy, $\left|\Phi_E(t)\right\rangle$ say, which depend on time only through the usual phase factor:

$$\left|\Phi_E(t)\right\rangle = \exp\left\{-iE(t-t_0)/\hbar\right\}\left|\Phi_E(t_0)\right\rangle. \tag{5.31}$$

These particular state vectors therefore simply rotate‡ with an angular frequency $\nu = E/\hbar$. It also follows that, for the energy eigenstates, the evolution operator may be written alternatively as

$$\mathsf{U}(t, t_0) = \exp\left\{-i\mathsf{H}(t-t_0)/\hbar\right\} \tag{5.32}$$

because, for any power of E, $E^n \to \mathsf{H}^n$ and hence, when working on an energy eigenstate, the exponentials in the last two equations are term-by-term equivalent.

The evolution operator (5.32) for the energy eigenstates of a time-independent Hamiltonian evidently satisfies the differential equation

$$i\hbar\frac{d}{dt}\mathsf{U}(t, t_0) = \mathsf{H}\mathsf{U}(t, t_0) \tag{5.33}$$

and is completely determined by the initial condition

$$\mathsf{U}(t_0, t_0) = \mathsf{I}. \tag{5.34}$$

It must now be asked whether these equations still determine the evolution operator when H contains the time and the state $\left|\Psi\right\rangle$ is arbitrary. That this is indeed the case follows at once on differentiating (5.30) and tentatively substituting (5.33); the result is simply (5.29), which is valid quite generally. The solution of (5.33), subject to initial condition (5.34), is thus the correct evolution operator; it is unique (first-order equation with one boundary condition) and corresponds to correct time development of $\left|\Psi\right\rangle$ at each instant.

---

‡The projection of $\left|\Phi_E(t)\right\rangle$ on a fixed vector $\left|\Phi_E(t_0)\right\rangle$ oscillates between $\pm\left|\Phi_E(t_0)\right\rangle$; the term "rotation" is used in a formal sense (cf. the vector diagrams used in discussing mechanical and electrical oscillations).

Equation (5.33) can be integrated formally, to give

$$U(t, t_0) = I - \frac{i}{\hbar} \int_{t_0}^{t} HU(t', t_0)\, dt' \qquad (5.35)$$

which is an integral equation whose explicit solution can be obtained by iteration (see, for example, Messiah (1961) Vol. II, Ch. 17). However, it is the *existence* of the time evolution operator U (and hence also its inverse $U^\dagger$) which concerns us here, rather than the methods for its determination. The actual use of the evolution operator is of great importance in the discussion of phenomena such as scattering.

We now return to (5.30) and note that in the Schrödinger picture all state vectors are evolving in time in a well-defined way. The essence of the so-called Heisenberg representation is the introduction of a unitary transformation which has the effect of reducing all vectors to rest; the states are then described by the time-independent kets $|\Psi(t_0)\rangle$ and the time-development of the system is thrown entirely into the operators representing the dynamical variables. This is of considerable advantage if one wishes to emphasize the connection between classical and quantum dynamics via the correspondence principle, which asserts that one description must merge into the other for $\hbar \rightarrow 0$. Noting that $U^\dagger$ exists (being the inverse operator leading from $\Psi(t)$ to $\Psi(t_0)$),

$$U^\dagger(t, t_0) = U(t_0, t). \qquad (5.36)$$

The Heisenberg and Schrödinger schemes (distinguished by subscripts H and S) are related by the standard transformation equations (p. 115). Hence

$$|\Psi_S\rangle = U|\Psi_H\rangle, \quad |\Psi_H\rangle = U^\dagger|\Psi_S\rangle. \qquad (5.37)$$

Operators in the new representation must be transformed in a corresponding way; on expressing any expectation value in terms of Heisenberg kets we obtain $\langle \Psi_S|A_S|\Psi_S\rangle = \langle \Psi_H|U^\dagger A_S U|\Psi_H\rangle$ and hence

$$A_H = U^\dagger A_S U. \qquad (5.38)$$

The Heisenberg operator is therefore time-dependent even when A_S is not.

To obtain the time-development of operators in the Heisenberg representation we differentiate (5.38) and obtain

$$\frac{dA_H}{dt} = \left(\frac{dU^\dagger}{dt}\right) A_S U + U^\dagger A_S \left(\frac{dU}{dt}\right) + U^\dagger \left(\frac{dA_S}{dt}\right) U.$$

But (5.33) gives‡

$$i\hbar \frac{d\mathsf{U}}{dt} = \mathsf{HU}, \quad -i\hbar \frac{d\mathsf{U}^\dagger}{dt} = \mathsf{U}^\dagger \mathsf{H}^\dagger$$

and substitution then yields

$$i\hbar \frac{d\mathsf{A}_H}{dt} = \mathsf{U}^\dagger[\mathsf{A}_S, \mathsf{H}]\mathsf{U} + i\hbar \mathsf{U}^\dagger \left(\frac{d\mathsf{A}_S}{dt}\right)\mathsf{U}. \tag{5.39}$$

Now H is defined in the Schrödinger picture: if we introduce a Heisenberg Hamiltonian

$$\mathsf{H}_H = \mathsf{U}^\dagger \mathsf{H} \mathsf{U} \tag{5.40}$$

and note also that

$$\mathsf{U}^\dagger \frac{d\mathsf{A}_S}{dt} \mathsf{U} = \frac{\partial \mathsf{A}_H}{\partial t}$$

(where the partial differentiation implies that we take account only of the explicit time-dependence in the observable as defined by A_S—not in the transformation operator U) it readily follows that

$$-\frac{\hbar}{i} \frac{d\mathsf{A}_H}{dt} = [\mathsf{A}_H, \mathsf{H}_H] - \frac{\hbar}{i} \frac{\partial \mathsf{A}_H}{\partial t} \tag{5.41}$$

where all operators are now in the Heisenberg representation. This is the *Heisenberg equation of motion*.

If there is no explicit time dependence in A_S, (5.41) reduces to

$$-\frac{\hbar}{i} \frac{d\mathsf{A}_H}{dt} = [\mathsf{A}_H, \mathsf{H}_H] \tag{5.42}$$

and if A_H and H_H commute the operator A_H becomes a constant of the motion (i.e. a time-independent *operator*). Since the states are described by time-independent kets it follows that a corresponding dynamical variable will also have a constant expectation value.

The Schrödinger and Heisenberg formulations are of course completely equivalent. The Schrödinger formulation is used almost exclusively in stationary state problems, and therefore in a large part of quantum chemistry. But the Heisenberg equations are often more convenient in the discussion of time-dependent phenomena and are therefore widely used in, for example, collision theory. Such developments are taken up in later Volumes.

---

‡Note that $(\mathsf{AB})^\dagger = \mathsf{B}^\dagger \mathsf{A}^\dagger$ and that in taking the adjoint the sign of i must be reversed (p. 53).

5.7.   Representation of incompletely specified states

So far, we have been concerned exclusively with a system characterized by some given state vector Ψ, for example an eigenstate of the
Hamiltonian operator H and of any other commuting operators whose
simultaneous eigenvalues define a state of maximal knowledge of the
system. Such a state vector is in general time-dependent, the dependence being particularly simple for a stationary state, and even in the
general case the time evolution of the state vector is determined in
principle by (4.1) and hence in terms of the evolution operator of
Section 5.6. But sometimes it is necessary to deal with systems whose
condition is not specified with the precision necessary for setting up a
state vector. It is important to note that a state of the type referred to
in framing Corollary 11 (p. 91) namely

$$\Psi = \sum c_K \Psi_K \tag{5.43}$$

where the Ψ_K are states of maximal knowledge and hence, for a
conservative system, eigenstates, is *not* "incompletely specified" in the
sense with which the phrase will now be used. The specification of
eigenstates and expansion coefficients uniquely defines the state vector
Ψ and its subsequent development: an incompletely specified state, on
the other hand, *has no unique state vector* and to describe it we shall have
to introduce the idea of an ensemble as used in statistical mechanics.

First we recall that if A is one of a maximal set of compatible
observables (p. 90 et seq.) with operator A, then the expectation value
in state (5.43) is

$$\langle A \rangle = \langle \Psi | A | \Psi \rangle = \sum_{K,L} c_L^* c_K \langle \Psi_L | A | \Psi_K \rangle \tag{5.44}$$

Although in Corollary 11 we have identified $|c_K|^2$ as a probability of
finding the system in a particular state Ψ_K of maximal knowledge (i.e.
an eigenstate of the maximal set of commuting operators to which A
belongs) we are referring to a system whose state vector Ψ is evolving
in such a way that the coefficients are fully determined. An *incompletely*
specified system, on the other hand, may be defined through an
expectation value expression of the form (5.44) but with a more general
weight factor which is not merely a product $c_K^* c_L$.

Let us take, for example, a system A (with states Ψ_1^A, Ψ_2^A, ...) in
weak interaction with its environment B which, in the terminology of
statistical thermodynamics, provides a "heat bath" (with states Ψ_1^B,
Ψ_2^B, ...). The wave function of $A + B$ can then be written

$$\Psi = \sum_{K,J} c_{KJ} \Psi_K^A \Psi_J^B$$

If our interest is in the system A, we obtain for the expectation value of any quantity Q associated with system A,

$$\langle Q \rangle = \langle \Psi | Q | \Psi \rangle = \sum_{L,M,K,J} c_{LM}{}^* c_{KJ} \langle \Psi_L{}^A | Q | \Psi_K{}^A \rangle \langle \Psi_M{}^B | \Psi_J{}^B \rangle$$

which may be written

$$\langle Q \rangle = \sum_{L,K} \rho_{KL} \langle \Psi_L{}^A | Q | \Psi_K{}^A \rangle \qquad (5.45)$$

This has the general form (5.44) but the condition of the system is characterized simply by the numerical coefficients $\rho_{KL} = \sum_J c_{KJ} c_{LJ}{}^*$ (using the orthonormality of the state vectors) and indeed *system A has no wave function of its own*. A concrete example of this kind is provided when A is an atom in its "valence state", resulting from "dissociation" of the molecule $A - B$ by a hypothetical reduction of the interaction between its constituent atoms, without change of wave function: it is sometimes said that the valence state is a "mixture of spectroscopic states" but this does not mean the state is expressible in the form (5.43)—there is no "coherence" between the spectroscopic states and the mixing implies only that expectation values appear as a suitable weighted mixture (5.45) of matrix elements associated with the spectroscopic states.

Ensembles

To allow in a general way for incomplete specification of states it is customary to set up a representative *ensemble*. The ensemble consists of a very large number (N) of hypothetical "copies" of the system of interest, system n being supposed to have a wave function

$$\Psi^{(n)} = \sum_n c_K{}^{(n)} \Psi_K \qquad (5.46)$$

and the *ensemble average* of an observable Q is then defined as the mean of the expectation values for the individual copies:

$$\langle Q \rangle = N^{-1} \sum_n \langle \Psi^{(n)} | Q | \Psi^{(n)} \rangle$$

$$= N^{-1} \sum_n \sum_{L,K} c_L{}^{(n)*} c_K{}^{(n)} \langle \Psi_L | Q | \Psi_K \rangle$$

which may be rewritten in the form

$$\langle Q \rangle = \sum_{K,L} \rho_{KL} \langle \Psi_L | Q | \Psi_K \rangle \qquad (5.47)$$

In other words the expectation value for an *incompletely* specified system (5.45) may be re-interpreted as the ensemble average expectation value for a large collection of copies in a variety of completely specified states. The coefficients ρ_{KL}, defined by averaging over the representative ensemble, are

$$\rho_{KL} = N^{-1}\sum_n c_K^{(n)} c_L^{(n)*} \tag{5.48}$$

and may be collected into a matrix ρ called the *density matrix*.

Since $\langle \Psi_L | Q | \Psi_K \rangle$ is the LK-element of the matrix associated with the operator Q in the representation provided by the basis $\{\Psi_K\}$, the result (5.47) may be written

$$\langle Q \rangle = \sum_{K,L} \rho_{KL} Q_{LK} = \text{tr } \rho Q = \text{tr } Q\rho \tag{5.49}$$

giving an elegant expression for the expectation value of any observable Q in a system specified (incompletely) only by a density matrix ρ with given elements ρ_{KL}.

A system in an incompletely specified state is thus described by a density matrix, not by a wave function, and the lack of precision in the specification is determined by the "spread" of states $\Psi^{(n)}$ admitted among the members of the ensemble—which is reflected in the specification of the elements ρ_{KL}. The important fact is that the expectation value of any quantity is now determined by *two* types of averaging—the first giving the usual quantum mechanical expectation value for a system in a given state, the second allowing for uncertainty in the specification of the state itself. If there is *no* uncertainty in the state, every $\Psi^{(n)}$ is the same and (5.48) reduces to $\rho_{KL} = c_K c_L^*$, exactly as for a single system; the system and its representative ensemble are then said to be in a *pure state*; otherwise the state is said to be *mixed*.

A complete mathematical justification of the choice of ensemble to represent a given physical situation is a matter of considerable difficulty, discussed in textbooks of statistical mechanics (e.g. Tolman, 1938; ter Haar, 1954, Appendix 1), and will not be considered here. We note, however, that if the $\{\Psi_K\}$ are energy eigenstates the choice

$$\rho_{KK} = c\exp\{-\beta E_K\}, \quad \rho_{KL} = 0 \quad (K \neq L) \tag{5.50}$$

will give a plausible representation of the state of a system in thermal equilibrium with a heat bath. In this case the expectation value of any observable Q is given by (5.47) as

$$\langle Q \rangle = \sum_K \rho_{KK} \langle \Psi_K | Q | \Psi_K \rangle = C e^{-\beta E_K} \langle \Psi_K | Q | \Psi_K \rangle \tag{5.51}$$

which indicates a Boltzman distribution of probabilities of finding the system in its possible energy eigenstates. This choice of density matrix elements would follow for an ensemble in which the $\Psi^{(n)}$ were taken to be the energy eigenstates themselves, with random phase factors; for the non-zero coefficients in (5.46) would then all be unimodular complex numbers and ρ_{KK} would give the fractional number of copies in state Ψ_K while ρ_{LK} would vanish in the summation over random phase factors. Ensembles defined in this way are described as "canonical" and play a fundamental rôle in equilibrium statistical mechanics.

The density matrix in Schrödinger language

We now develop the idea of the density matrix independently from the standpoint of a particular quantum mechanical language. It is convenient to start from the Schrödinger language and to define

$$\rho(\mathbf{x};\mathbf{x}') = \Psi(\mathbf{x})\Psi^*(\mathbf{x}') \tag{5.52}$$

as the Schrödinger density matrix for a pure state in which the system is definitely in state Ψ (which may or may not be a stationary state, though the time variable will not be shown explicitly). The name "matrix" is used in this context because the variables $\mathbf{x}$ and $\mathbf{x}'$ (indicating, as usual, *all* variables collectively) correspond formally to the row and column indices in the matrix ρ_{KL} exactly as in Section 5.3. We shall find presently that $\rho(\mathbf{x};\mathbf{x}')$ and ρ_{KL} are in fact simply alternative representations of a density *operator* ρ.

The expectation value of an operator $\mathbf{Q}$ in the pure state Ψ, may then be written

$$\langle Q \rangle = \int \Psi^*(\mathbf{x})\mathbf{Q}\Psi(\mathbf{x})\,d\mathbf{x}$$
$$= \int [\mathbf{Q}\Psi(\mathbf{x})\Psi^*(\mathbf{x}')]_{\mathbf{x}'\to\mathbf{x}}\,d\mathbf{x}$$

where the variables in Ψ^* have been primed in order to "protect" Ψ^* from the operator, which by convention operates on the unprimed variables $\mathbf{x}$ (and hence upon Ψ only), and $\mathbf{x}'$ is identified with $\mathbf{x}$ after operating with $\mathbf{Q}$. The expectation value thus becomes

$$\langle Q \rangle = \int [\mathbf{Q}\rho(\mathbf{x};\mathbf{x}')]_{\mathbf{x}'\to\mathbf{x}}\,d\mathbf{x} \tag{5.53}$$

This operation corresponds formally to "summing over the diagonal elements" in matrix theory, and is therefore formally similar to taking the trace of $\mathbf{Q}\rho$ in the matrix equation (5.49).

Now consider the more general quantity

$$\rho(\mathbf{x};\mathbf{x}') = \sum_{K,L} \rho_{KL}\Psi_K(\mathbf{x})\Psi_L^*(\mathbf{x}') \tag{5.54}$$

where ρ_{KL} are arbitrary numerical coefficients and Ψ_K, Ψ_L belong to any complete orthonormal set for the system considered. This more general density matrix coincides with the pure state form (5.52) only in very special cases, when the ρ_{KL} are products of expansion coefficients $(c_K c_L^*)$ arising from a wavefunction of the form (5.43); otherwise, as we now show, it describes the mixed state in which ρ_{KL} can be regarded as the ensemble average defined in (5.48).

To demonstrate this equivalence we show first that (5.53), with $\rho(\mathbf{x};\mathbf{x}')$ defined generally in (5.54), leads to exactly the same expectation value formula as we obtained from the ensemble approach. Thus

$$\langle Q \rangle = \int [\sum_{K,L} \rho_{KL} Q \Psi_K(\mathbf{x}) \Psi_L{}^*(\mathbf{x}')]_{x' \to x} d\mathbf{x} = \sum_{K,L} \rho_{KL} \langle \Psi_L | Q | \Psi_K \rangle$$

which coincides with (5.47). Secondly, we show, using the ortho-normality of the expansion functions, that ρ_{KL} is simply the KL matrix element of a *density operator* which is represented in Schrödinger language by an integral operator with kernel $\rho(\mathbf{x};\mathbf{x}')$. Thus, by definition, the PQ element of ρ is

$$\langle \Psi_P | \rho | \Psi_Q \rangle = \int \Psi_P{}^*(\mathbf{x}) \rho(\mathbf{x};\mathbf{x}') \Psi_Q(\mathbf{x}') d\mathbf{x}' d\mathbf{x}$$

and on inserting (5.54) and completing the integrations we obtain

$$\langle \Psi_P | \rho | \Psi_Q \rangle = \rho_{PQ} \qquad (5.55)$$

It is now evident that (5.53) and (5.47) are entirely equivalent state-ments, the first in a continuous representation, using the language of integral operators, the second in a discrete representation, in which each operator is represented by a matrix: the trace formula (5.49) becomes generally valid if the "trace" in a continuous representation is interpreted as an integration over "diagonal elements" according to (5.53). It is also clear that the results may easily be transposed into the language provided by any other orthonormal basis $\{\Psi_K'\}$: for if $\Psi_K' = \sum_L \Psi_L U_{LK}$, where the U_{LK} form a unitary matrix, then

$$\rho \to \rho' = \mathbf{U}^\dagger \rho \mathbf{U}, \quad Q \to Q' = \mathbf{U}^\dagger Q \mathbf{U} \qquad (5.56)$$

exactly as in (3.69), and consequently‡

$$\langle Q \rangle = \operatorname{tr} Q\rho = \operatorname{tr} Q'\rho'$$

giving invariance against change of quantum mechanical language. The fact that ρ_{KL} and $\rho(\mathbf{x};\mathbf{x}')$ are simply alternative representations of the same *density operator* ρ is particularly clear when Dirac notation is employed. Thus (5.54) could be written, with the abbreviation $|\Psi_K\rangle = |K\rangle$,

$$\rho(\mathbf{x};\mathbf{x}') = \sum_{K,L} \langle \mathbf{x} | K \rangle \rho_{KL} \langle L | \mathbf{x}' \rangle$$

or, since by (5.55) $\rho_{KL} = \langle K | \rho | L \rangle$,

---

‡Thus, $\operatorname{tr} Q'\rho' = \operatorname{tr} \mathbf{U}^\dagger Q \mathbf{U} \mathbf{U}^\dagger \rho \mathbf{U} = \operatorname{tr} \mathbf{U}^\dagger Q \rho \mathbf{U} = \operatorname{tr} Q\rho$ (invariance of the trace under a cyclic permutation).

$$\rho(\mathbf{x};\mathbf{x}') = \sum_{K,L} \langle\mathbf{x}|K\rangle\langle K|\rho|L\rangle\langle L|\mathbf{x}'\rangle = \langle\mathbf{x}|\rho|\mathbf{x}'\rangle \qquad (5.57)$$

in which $\Psi_K(\mathbf{x}) = \langle\mathbf{x}|K\rangle$ and $\Psi_L^*(\mathbf{x}') = \langle L|\mathbf{x}'\rangle$ play the part of transformation functions connecting discrete and continuous representations, and $\langle\mathbf{x}|\rho|\mathbf{x}'\rangle$ indicates an alternative notation for the integral kernel $\rho(\mathbf{x};\mathbf{x}')$. The density operator may thus be written formally as

$$\rho = \sum_{K,L} |K\rangle\langle K|\rho|L\rangle\langle L| = \sum_{K,L} \rho_{KL}|K\rangle\langle L| \qquad (5.58)$$

i.e. as a sum of ket-bra operators. Such an operator is characterized by its effect on any ket $|\rangle$. It produces a new ket $|\rangle'$ according to

$$|\rangle' = |K\rangle\langle L|\rangle \qquad (5.59)$$

which is simply $|K\rangle$ multiplied by a number, the scalar product $\langle L|\rangle$. A ket-bra product is thus a generalized projection operator of "dyad" (Appendix 4). In a continuous representation it is described by the integral kernel

$$\langle\mathbf{x}|K\rangle\langle L|\mathbf{x}'\rangle$$

whose effect on the ket represented by $\langle\mathbf{x}|\rangle$ is to produce

$$\langle\mathbf{x}|\rangle' = \int\langle\mathbf{x}|K\rangle\langle L|\mathbf{x}'\rangle\langle\mathbf{x}'|\rangle d\mathbf{x}' = \langle\mathbf{x}|K\rangle\langle L|\rangle \qquad (5.60)$$

or in wave function language

$$\Psi'(\mathbf{x}) = \int\Psi_K(\mathbf{x})\Psi_L^*(\mathbf{x}')\Psi(\mathbf{x}')d\mathbf{x}' = \Psi_K(\mathbf{x})\langle L|\rangle.$$

The last two equations simply express the symbolic result (5.59) in Schrödinger language.

Properties of the density matrix

The matrix ρ is, from its definition (5.48), clearly Hermitian; $\rho_{LK} = \rho_{KL}^*$. A new basis may therefore always be chosen so that the density matrix becomes diagonal and for the moment we assume, without loss of generality, that ρ has this form. The first obvious conclusions are that

$$\mathrm{Tr}\,\rho = \sum_K \rho_{KK} = 1 \qquad (a) \qquad (5.59)$$

$$0 \leqslant \rho_{KK} \leqslant 1 \qquad (b)$$

which follow immediately from the fact that ρ_{KK} is the fractional number of ensemble members in state Ψ_K. Since the trace is invariant under the unitary transformation (5.56), the normalization (5.59a) is independent of the particular representation used. We may then infer that $\mathrm{Tr}\,\rho^2 \leqslant 1$ and that consequently, even when ρ is non-diagonal,

$$\sum_{K,L} \rho_{KL}\rho_{LK} = \sum_{K,L} |\rho_{KL}|^2 \leqslant 1 \tag{5.60}$$

showing that *none* of the individual elements, in any representation, can exceed unity.

Next we ask what property characterizes the density matrix representing a system in a *pure* state with wave function Ψ. In this case every copy in the ensemble has the same wave function

$$\Psi^{(n)} = \Psi = \sum_K c_K \Psi_K \tag{5.61}$$

and (5.48) gives $\rho_{KL} = c_K c_L^*$. The normalization condition $\sum_L c_L c_L^* = 1$ then implies a density matrix condition:

$$\sum_L \rho_{KL}\rho_{LM} = \sum_L c_K c_L^* c_L c_M^* = c_K c_M^* = \rho_{KM}.$$

In other words the density matrix must be *idempotent*,

$$\rho^2 = \rho \tag{5.62}$$

This is also a *sufficient* condition to characterize a pure state: for in a diagonal representation (5.62) implies that

$$\rho_{KK} = 0 \text{ or } 1 \quad (\text{all } K).$$

But since $\rho_{KK} = |c_K|^2$, and the wave function (5.61) is normalized, there can be *only one non-zero coefficient*; if this corresponds to $\rho_{KK} = 1$ then the corresponding basis function Ψ_K, whose phase is arbitrary, represents the pure state of the system. This important result also follows immediately from the definition in Schrödinger language; the pure state density matrix (5.52) is idempotent in the sense

$$\int \rho(\mathbf{x};\mathbf{x}'')\rho(\mathbf{x}'';\mathbf{x}')d\mathbf{x}'' = \int \Psi(\mathbf{x})\Psi^*(\mathbf{x}'')\Psi(\mathbf{x}'')\Psi^*(\mathbf{x}')d\mathbf{x}''$$

$$= \Psi(\mathbf{x})\Psi^*(\mathbf{x}') = \rho(\mathbf{x};\mathbf{x}') \tag{5.63}$$

and the density operator which it describes is therefore a projection operator. There is one such projection operator for each state of a complete set $\{\Psi_K\}$; and if Ψ_K is an eigenfunction with eigenvalue 1, then Ψ_L (all $L \neq K$) are eigenfunctions with eigenvalue 0. The idempotency condition on the matrix (5.62), or on the kernel (5.63), is thus basically a requirement that the density *operator* shall be a projection operator onto a single pure state.

Finally, to complete the generalization of quantum mechanics to incompletely specified states, we require the analogues of Schrödinger's equations: both the stationary and time dependent equations follow if we consider the time development of the density matrix defined by (5.46) and (5.48).

The most general state $\Psi^{(n)}$, of the nth copy in the ensemble, is given by (5.46) in terms of the complete set Ψ_K. The time development of the coefficients follows on substituting the expansion in (4.1), multiplying by Ψ_K^* and integrating over all variables $(\mathbf{x})$, the result being,

$$\frac{\partial c_K^{(n)}}{\partial t} = -\frac{i}{\hbar} \sum_L H_{KL} c_L^{(n)}. \tag{5.64}$$

We then obtain, using the shorthand notation $\mathbf{c}^{(n)} = -(i/\hbar)\mathbf{Hc}^{(n)}$ and noting that $c_K^{(n)} c_L^{(n)*} = [\mathbf{c}^{(n)}\mathbf{c}^{(n)\dagger}]_{KL}$,

$$\frac{\partial}{\partial t}[\mathbf{c}^{(n)}\mathbf{c}^{(n)\dagger}] = -\frac{i}{\hbar}\{\mathbf{Hc}^{(n)}\mathbf{c}^{(n)\dagger} - \mathbf{c}^{(n)}(\mathbf{Hc}^{(n)})^{\dagger}\}.$$

On writing $(\mathbf{Hc}^{(n)})^{\dagger} = \mathbf{c}^{(n)\dagger}\mathbf{H}^{\dagger} = \mathbf{c}^{(n)\dagger}\mathbf{H}$ and averaging over the ensemble this yields (since the average of $c_K^{(n)} c_L^{(n)*}$ is ρ_{KL})

$$\frac{\partial \rho}{\partial t} = -\frac{i}{\hbar}\{\mathbf{H}\rho - \rho\mathbf{H}\}. \tag{5.65}$$

This result is valid in any language and completely determines the time development of the density matrix describing any system, completely or incompletely specified; it is thus the generalization of the time-dependent Schrödinger equation.

Stationary states are now defined as those for which $\partial \rho / \partial t = 0$ and, consequently, all time-independent operators have constant expectation values: such states are thus determined by the commutation condition

$$\mathbf{H}\rho - \rho\mathbf{H} = 0. \tag{5.66}$$

This equation shows that, in general, stationary states may exist even when the system of interest is in an incompletely specified condition; it is satisfied, for instance, for a system described by the canonical ensemble with density matrix (5.50) as may be seen by regarding the Ψ_K as energy eigenstates, in which case $\mathbf{H}$ and ρ become simultaneously diagonal. Such states, in which an energy uncertainty remains, are appropriate in conditions of thermal equilibrium. Pure states of the ensemble, corresponding to a system definitely known to be in an eigenstate Ψ_K with energy E_K are distinguished by density matrices satisfying the further condition (5.62). Since idempotency requires that ρ in its diagonal form has only one non-zero element, ρ_{KK} say, (5.66) becomes

$$\sum_M (H_{LM}\rho_{MN} - \rho_{LM}H_{MN}) = \delta_{NK}H_{LK}\rho_{KM} - \delta_{LK}\rho_{LK}H_{KN} = 0.$$

Thus $H_{LK} = H_{KL} = 0$ $(L \neq K)$ and this diagonality condition requires that each Ψ_K is an eigenfunction of H. We thus retrieve the time-independent Schrödinger equation

$$H\Psi = E\Psi \tag{5.67}$$

as a condition for a stationary state of a *pure* ensemble.

Apart from its obvious importance in statistical mechanics (e.g. ter Haar, 1954; Mayer 1968) density matrix theory has important implications in many-particle quantum mechanics (Vol. 3). For further developments in this field the reader is referred to the literature and particularly to the review articles (see, for example, ter Haar, 1961; Fano, 1957; Löwdin, 1955; McWeeny, 1960).

REFERENCES

COULSON, C. A. (1941) *Proc. Camb. Phil. Soc.* **37**, 55 (and subsequent papers with Duncanson).

DIRAC, P. A. M. (1947) *The Principles of Quantum Mechanics*, 3rd ed., Clarendon Press, Oxford.

EPSTEIN, I. R. (1970) *J. Chem. Phys.* **52**, 3838.

FANO, U. (1957) *Rev. Mod. Phys.* **29**, 74.

HENNEKER, W. H. and CADE, P. E. (1968) *Chem. Phys. Letters* **2**, 575.

LÖWDIN, P.-O. (1955) *Phys. Rev.* **97**, 1474.

McWEENY, R. (1960) *Rev. Mod. Phys.* **32**, 335.

MESSIAH, A. (1961) *Quantum Mechanics*, Vols. I and II, North Holland, Amsterdam.

PAULING, L. and PODOLSKY, G. (1929) *Phys. Rev.* **34**, 109.

TER HAAR, D. (1954) *Elements of Statistical Mechanics*, Rinehart, Holt, Winston, New York.

TER HAAR, D. (1961) *Rep. Prog. Phys.* **24**, 304.

TOLMAN, R. C. (1938) *The Principles of Statistical Mechanics*, Oxford University Press, London.

APPENDIX 1

THE SCHRÖDINGER EQUATION IN GENERALIZED COORDINATES

For many purposes it is expedient to replace the Cartesian coordinates and their associated momenta by "generalized coordinates" better suited to the form of the system; in a system with spherical symmetry, for example, it would be natural to introduce spherical polar coordinates.

The basic property of the generalized coordinates is that the kinetic energy may be expressed as a quadratic form in their time derivatives:

$$T = \tfrac{1}{2}\sum_{i,k} M_{ik}\dot{q}_i\dot{q}_k \tag{A1.1}$$

where the coefficients M_{ik} are time-independent functions of position, depending on choice of coordinate system. The corresponding generalized momenta, $p_1, p_2, \ldots, p_n$ are then defined by

$$p_i = \frac{\partial T}{\partial \dot{q}_i}. \tag{A1.2}$$

Thus in the Cartesian case

$$T = \tfrac{1}{2}m \sum_{i=1}^{3} \dot{q}_i{}^2 \quad (q_i = x, y, z; M_{ik} = m\,\delta_{ik})$$

and it follows at once that $p_i = m\dot{q}_i$.

In Chapter 1, quantum mechanical operators were associated with the momenta and the kinetic energy, *expressed in Cartesian coordinates*, according to

$$p_i \to \mathsf{p}_i = \frac{\hbar}{i}\frac{\partial}{\partial q_i}, \tag{A1.3}$$

$$H(q_1 \ldots q_n\, p_1 \ldots p_n) \to \mathsf{H}(q_1 \ldots q_n\, \mathsf{p}_1 \ldots \mathsf{p}_n). \tag{A1.4}$$

We now wish to make this association in such a way that it can be carried over at once from one coordinate system to another. This is accomplished most elegantly and generally by tensor methods; here we simply explain the principles involved. T is an *invariant*, having the same value for a given dynamical situation, irrespective of the coordinate system used. The $\dot{q}_i$, however, transform in a characteristic way when the coordinates are changed, for $\bar{q}_i = \bar{q}_i(q_1, q_2, \ldots q_n)$ implies

$$d\bar{q}_i = \sum_j \left(\frac{\partial \bar{q}_i}{\partial q_j}\right) dq_j \qquad (A1.5)$$

and the infinitesimals (and consequently the $\dot{q}_i$) are then said to be "contravariant components of a tensor of rank 1". In order that T shall be invariant, the coefficients M_{ik} in (A1.1) must transform in a "reciprocal" fashion; they are "covariant components of a tensor of rank 2". Similarly, the momentum components defined by (A1.2) transform like covariant components of rank 1; and so do the associated operators defined in (A1.3). The association (A1.3) is therefore independent of the coordinate system employed; the operators and the momenta with which they are associated follow the same transformation law and, if (A1.3) is valid in a Cartesian system, it will be valid always.

The operator associated with T, however, is more difficult since (A1.1) as it stands does not contain the p's. If we use a Cartesian system, we can, of course, obtain $T = (1/2m)\sum_i p_i{}^2$; but if p_i is then interpreted according to (A1.3) the resultant operator will be valid *only* for Cartesian systems and will not remain invariant on changing to a more general system. What we need is a general expression for T in terms of the p's which, on making the substitution (A1.3), will give an *invariant differential operator* T. This must reduce to the usual sum of second derivatives in the Cartesian case, but must remain valid on introducing *any* other coordinates $\bar{q}_i = \bar{q}_i(q_1, q_2, \ldots q_n)$.

The required expression for T is found to be

$$T = \tfrac{1}{2}\frac{1}{\sqrt{G}} \sum_{i,k} p_i(\sqrt{G}\, M^{ik}) p_k, \qquad (A1.6)$$

where, using **M** for the matrix of coefficients M_{ik},

$$G = \det \mathbf{M}, \quad M^{ik} = (\mathbf{M}^{-1})_{ik}. \qquad (A1.7)$$

Consequently, M^{ik} is the cofactor of M_{ik} in G, divided by the determinant itself. On using the invariant form (A1.6), a corresponding operator T may be set up by the standard association (A1.3). The Schrödinger equation then becomes

$$\mathsf{H}\Psi = -\frac{\hbar^2}{2}\frac{1}{\sqrt{G}} \sum_{i,k} \frac{\partial}{\partial q_i}\left(\sqrt{G}\, M^{ik} \frac{\partial \Psi}{\partial q_k}\right) + V\Psi = E\Psi \qquad (A1.8)$$

which is completely general.

If we consider the special case of a one-particle system, we may infer from (A1.8) the most general form of the operator ∇^2 in an arbitrary coordinate system. More directly, however, we start from the expression for the square of the distance between two points in the general form (cf. A1.1)

$$ds^2 = \sum_{i,k} m_{ik} dq_i dq_k. \tag{A1.9}$$

This is the basic invariant, whose form in any given coordinate system is obtained by purely geometrical considerations. The invariant expression for ∇^2 takes the special form $\sum_i (\partial/\partial x_i)(\partial/\partial x_i)$ in a Cartesian system, but the operators (like the p_i) transform covariantly under change of coordinates and the general form becomes (cf. A1.6)

$$\nabla^2 = \frac{1}{\sqrt{g}} \sum_{i,k} \frac{\partial}{\partial q_i} \left(\sqrt{g}\, m^{ik} \frac{\partial}{\partial q_k} \right), \tag{A1.10}$$

where, using **m** for the matrix of coefficients in the invariant (A1.9),

$$g = \det \mathbf{m}, \quad m^{ik} = (\mathbf{m}^{-1})_{ik} \tag{A1.11}$$

as in (A1.8). The essential difference between the forms occurring in (A1.8) and (A1.10) is that the latter depends only on geometry of the system while the former includes particle masses.

Finally, it must be noted that in integrations over all space the volume element, whose form is immediate in a Cartesian system, must also be written in such a way as to ensure invariance of the integrand. The appropriate element for volume integration turns out to be

$$d\mathbf{q} = \sqrt{g}\, dq_1 dq_2 \ldots dq_n \tag{A1.12}$$

where g is the determinant introduced in (A1.11).

There is one very important three-dimensional case of the above results; this occurs when the volume element, bounded by neighbouring coordinate surfaces on which q_1, q_2 and q_3, respectively, are constant, is *rectangular* (Fig. A1.1). If we suppose the surfaces corresponding to

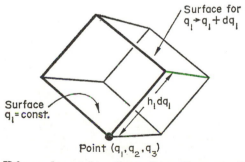

FIG. A1.1. Volume element in orthogonal curvilinear coordinates. The element is rectangular, opposite faces corresponding to change dq_i in one of the variables (other variables held constant). See also Fig. A2.1.

values q_i and $q_i + dq_i$ (other q's constant) are separated by a distance $h_i dq_i$, then the basic form (A1.9) becomes

$$ds^2 = h_1{}^2 dq_1{}^2 + h_2{}^2 dq_2{}^2 + h_3{}^2 dq_3{}^2 \qquad (A1.13)$$

and all results are expressible in terms of the three *scale factors* h_1, h_2, h_3. The matrix **m** has diagonal elements (all positive) $m_{ii} = h_i{}^2$; the determinant g is $g = h_1{}^2 h_2{}^2 h_3{}^2$ and the coefficients m^{ii} easily follow as $m^{ii} = 1/h_i{}^2$. The volume element and the Laplacian (A1.10) then take the forms

$$d\mathbf{q} = h_1 h_2 h_3 dq_1 dq_2 dq_3, \qquad (A1.14)$$

$$\nabla^2 = \frac{1}{h_1 h_2 h_3} \sum_i \frac{\partial}{\partial q_i} \left(\frac{h_1 h_2 h_3}{h_i{}^2} \frac{\partial}{\partial q_i} \right). \qquad (A1.15)$$

The special form (A1.15) is more widely useful than (A1.10) or (A1.8), which are required in relatively few applications (see, for example, Wilson, Decius and Cross, 1955, Ch. 4). A derivation of (A1.10) may be found in the book by Margenau and Murphy (1943), Section 5.17.

REFERENCES

WILSON, E. B., DECIUS, J. C. and CROSS, P. C. (1955) *Molecular Vibrations*, McGraw Hill, New York.
MARGENAU H. and MURPHY, G. M. *The Mathematics of Physics and Chemistry*, Van Nostrand, Toronto, New York, London.

SEPARATION OF
PARTIAL DIFFERENTIAL EQUATIONS

THE Schrödinger equation is a partial differential equation, and when solutions can be obtained in closed form it is normally a result of *separating the variables*. Separability depends on the use of a suitable coordinate system; the only choices for which separation of the Schrödinger equation has proved possible are the eleven possible *orthogonal* coordinate systems (Eisenhart, 1934) in which ∇^2 takes the form (A1.15). Here we give one example to illustrate the separation technique, before stating the general criteria for separability.

EXAMPLE. *Cylindrical coordinates.* Let us consider the Schrödinger equation for an electron moving in a potential field, of axial symmetry, constant in the z direction in the region between infinite walls at $z = 0, d$. V is thus a function of distance from the axis (ρ, say):

$$V(r) = F(\rho) \quad (0 < z < d), \quad = \infty \text{ (otherwise)}.$$

It seems natural to adopt ρ, θ and z as coordinates; and in this case the volume element (Fig. A2.1) has sides of length

$$d\rho, \quad \rho d\theta, \quad dz.$$

The scale factors are thus $h_1 = h_3 = 1$, $h_2 = \rho$. On inserting these values in (A1.15) the Schrödinger equation with the assumed potential energy function takes the form

$$-\frac{\hbar_2}{2m} \frac{1}{\rho} \left[\frac{\partial}{\partial \rho} \left(\rho \frac{\partial \phi}{\partial \rho} \right) + \frac{\partial}{\partial \theta} \left(\frac{1}{\rho} \frac{\partial \phi}{\partial \theta} \right) + \frac{\partial}{\partial z} \left(\rho \frac{\partial \phi}{\partial z} \right) \right] + F(\rho) \phi = E \phi.$$

In a solution of separated form, ϕ will contain a factor depending only on *one* of the three variables, satisfying an ordinary differential equation in this variable alone. As we noted in section 2.1 such an equation will arise if, possibly after multiplication by some suitable factor, some terms of the partial differential equation become independent of all coordinates but one—this one coordinate not appearing elsewhere.

Let us try to find a solution of the form

$$\phi(\rho, \theta, z) = u(\rho, \theta)v(z).$$

135

On inserting this form we obtain

$$-\frac{\hbar^2}{2m}\left[\frac{v}{\rho}\frac{\partial}{\partial\rho}\left(\rho\frac{\partial u}{\partial\rho}\right)+\frac{v}{\rho^2}\frac{\partial^2 u}{\partial\theta^2}+u\frac{\partial^2 v}{\partial z^2}\right]+F(\rho)\phi=E\phi$$

and it is clear that separation is achieved on dividing by $\phi(=uv)$; for then

$$-\frac{\hbar^2}{2m}\left\{\frac{1}{\rho u}\frac{\partial}{\partial\rho}\left(\rho\frac{\partial}{\partial\rho}\right)+\frac{1}{\rho^2 u}\frac{\partial^2 u}{\partial\theta^2}\right\}+F(\rho)+\left[-\frac{\hbar^2}{2m}\frac{1}{v}\frac{\partial^2 v}{\partial z^2}\right]=E.$$

The term in square brackets depends only on z, which does not appear elsewhere, and (by the argument of p. 24) may be equated to a constant (E_z, say). The other terms on the left may likewise be equated to a constant $E_{\rho\theta}$, such that $E = E_{\rho\theta} + E_z$. This procedure gives

$$(\partial^2 v/\partial z^2) = -(2m/\hbar^2)E_z v$$

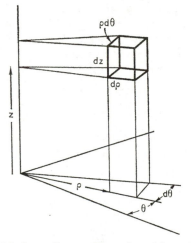

FIG. A2.1. Cylindrical coordinates. The volume element is rectangular, with sides of length $d\rho$, $\rho d\theta$, dz.

to determine the v factor in the wave function. The general solution

$$v(z) = A\exp(iz\sqrt{(2mE_z)}/\hbar)+B\exp(-iz\sqrt{2mE_z}/\hbar)$$

satisfies the boundary conditions $\phi = 0$ for $z = 0$, d when $B = -A$ and $\sin(d\sqrt{(2mE_z)}/\hbar) = 0$. The appropriate v-factor is thus

$$v(z) = \sin(n\pi z/d),$$

where n is an integral quantum number, restricting the separation constant to the values

$$E_z = n^2\pi^2\hbar^2/2md^2.$$

There remains an equation in *two* variables for the remaining factor $u(\rho, \theta)$. It is left as an exercise for the reader to show that the u-equation may be separated by a second application of the method.

The above Example suggests that separation of the Schrödinger equation will be possible only when the scale factors depend upon the coordinates in a simple way; the most general dependence admissible is in fact

$$h_2 h_3 / h_1 = f_1(q_1) F_1(q_2, q_3), \text{ etc.,} \tag{A2.1}$$

where the other two relations follow by cyclic permutation of subscripts 1, 2, 3. If we insert (A2.1) and (A1.15) in the Schrödinger equation (1.24), and postulate a completely separated form

$$\phi(q_1, q_2, q_3) = Q_1(q_1) Q_2(q_2) Q_3(q_3), \tag{A2.2}$$

the equation becomes, on dividing throughout by ϕ,

$$-\frac{\hbar^2}{2m} \left\{ \frac{F_1(q_2, q_3)}{Q_1(q_1)} \frac{\partial}{\partial q_1} [f_1(q_1) Q_1(q_1)] + \frac{F_2(q_3, q_1)}{Q_2(q_2)} \frac{\partial}{\partial q_2} [f_2(q_2) Q_2(q_2)] \right.$$

$$\left. + \frac{F_3(q_1, q_2)}{Q_3(q_3)} \frac{\partial}{\partial q_3} [f_3(q_3) Q_3(q_3)] \right\} + V(q_1, q_2, q_3) = E. \tag{A2.3}$$

The condition for separability may be stated as follows: *Equation* (A2.3) *can be separated if we can find a multiplying factor which will make at least one term depend on one coordinate only, this coordinate not appearing elsewhere.* Suppose, for example, we can find such a function $g_1(q_2, q_3)$ which on multiplication reduces (A2.3) to the form

$$-\frac{\hbar^2}{2m} \left\{ G_1(q_1) \frac{\partial}{\partial q_1} [f_1(q_1) Q_1(q_1)] + G_2(q_2, q_3] \frac{\partial}{\partial q_2} [f_2(q_2) Q_2(q_2)] \right.$$

$$\left. + G_3(q_2, q_3) \frac{\partial}{\partial q_3} [f_3(q_3) Q_3(q_3)] \right\} + U_1(q_1) + U_2(q_2, q_3) = E g_1(q_2, q_3).$$

$$\tag{A2.4}$$

Then the term

$$-\frac{\hbar^2}{2m} G_1(q_1) \frac{\partial}{\partial q_1} [f_1(q_1) Q_1(q_1)] + U_1(q_1)$$

depends only on the coordinate q_1, which appears nowhere else in the equation. Consequently this term must be a *constant*, a separation constant, and the resultant q_1 equation may be solved as an ordinary differential equation. We have already observed that such an equation may possess satisfactory solutions (e.g. quadratically integrable, or satisfying given boundary conditions) only for certain special values of the constant. When such a value is inserted back in (A2.4) there remains an equation in *two* variables (q_2, q_3) and we may attempt to separate this equation in its turn in a precisely similar way.

Other examples of the separation technique occur elsewhere (e.g. Chap. 2 of Vol. 2). Here we only stress that the possibility of separation depends on the nature of the potential function and the forms of any boundaries there may be. If each equipotential surface or boundary is specified by constancy of a certain coordinate, for example q_1 (with values of q_2, q_3 defining all points on the surface $q_1 =$ constant), then this particular choice of coordinates will facilitate separation. Thus, in the example opening this Appendix, the equipotentials were cylindrical, V depending only on the axial distance (ρ) within a region bounded by planes at $z = 0$, d; thus ρ ($= q_1$) and z ($= q_3$) were dictated by the geometry of the system, the third coordinate ($q_2 = \theta$) then giving the required rectangular volume element.

REFERENCE

EISENHART, L. P. (1934) *Phys. Rev.* **45**, 427.

APPENDIX 3

SERIES SOLUTION OF SECOND-ORDER DIFFERENTIAL EQUATIONS

THE differential equations encountered in Chapters 2 and 3, and in many other applications of quantum mechanics, are of the form

$$\frac{d^2y}{dx^2} + P(x)\frac{dy}{dx} + Q(x)y = 0 \qquad (A3.1)$$

and, since closed-form solutions are rarely obtainable, we proceed at once to solution in series. We suppose that $y = a_0$ at some convenient point $x = x_0$ and make a Taylor expansion about this point:

$$y = f(x_0+h) = f(x_0) + hf'(x_0) + (h^2/2!)f''(x_0) + \ldots \qquad (A3.2)$$

$$= a_0 + a_1(x-x_0) + a_2(x-x_0)^2 + \ldots$$

The validity of such an expansion depends on y being single valued, with all its derivatives existing, at $x = x_0$: the function $y = f(x)$ is then said to be *analytic* at $x = x_0$. Since higher derivatives can be obtained by further differentiation of (A3.1), the implication is that $P(x)$ and $Q(x)$ must also be analytic at $x = x_0$. We therefore distinguish two kinds of point:

> If $P(x)$ and $Q(x)$ are analytic at the point $x = x_0$, this point is an *ordinary point* of the differential equation (A3.1); otherwise it is a *singular point*. The second and higher derivatives are determinate at an ordinary point but not at a singular point. (A3.3)

A series solution of the form (A3.2) can always be obtained about an ordinary point. We give one illustration:

EXAMPLE 1. Consider $(d^2y/dx^2) + xy = 0$ (which has no solution in terms of elementary functions). Clearly $x = 0$ is an ordinary point, and d^2y/dx^2 is determinate $(= 0)$, so an expansion (A3.2) is valid.

Thus we obtain

$$(d^3y/dx^3) = -x(dy/dx) - y,$$

$$(d^4y/dx^4) = -x(d^2y/dx^2) - 2(dy/dx),$$

etc.

If we suppose $y = y_0$ and $(dy/dx) = y_0'$ at $x = x_0$ we then have $(d^2y/dx^2)_0 = 0$, $(d^3y/dx^3)_0 = -y_0$, $(d^4y/dx^4)_0 = -2y_0'$, etc., and obtain the Taylor series

$$y = y_0 + x \left(\frac{dy}{dx}\right)_0 + \frac{1}{2!} x^2 \left(\frac{d^2y}{dx^2}\right)_0 + \cdots$$

in the explicit form

$$y = y_0 \left(1 - \frac{x^3}{3!} + \frac{4x^6}{6!} - \cdots\right) + y_0' \left(x - \frac{2x^4}{4!} + \frac{10x^7}{7!} - \cdots\right).$$

This is in fact a general solution since y_0 and y_0' may be regarded as arbitrary constants (the full number for a *second*-order equation).

Instead of using the Taylor expansion directly, as in the above Example, it is often easier simply to insert the form (A3.2) into (A3.1) and then determine the coefficients a_0, a_1, a_2, ... by equating to zero the coefficient of each power of x. In this way it is sometimes possible to obtain a general relationship between successive coefficients from which the infinite series is easily generated. The last example may be treated in this way:

EXAMPLE 2. On substituting $y = a_0 + a_1 x + a_2 x^2 + \ldots$ into the differential equation $(d^2y/dx^2) + xy = 0$ we obtain $2a_2 + x(a_0 + 3.2a_3) + x^2(a_1 + 4.3a_4) + x^3(a_2 + 5.4a_5) + \ldots = 0$. Equating to zero the coefficient of each power of x we see that a_0 and a_1 are undetermined but that $a_2 = 0$ and subsequent coefficients are related in the following way:

$$a_{n-3} + n(n-1)a_n = 0.$$

Hence we obtain

$$a_3 = -a_0/3.2, \quad a_6 = -a_3/6.5 = 4a_0/6!, \quad \text{etc.,}$$

and

$$a_4 = -a_1/4.3, \quad a_7 = -a_4/7.6 = 10a_1/7!, \quad \text{etc.}$$

There are thus two distinct sets of coefficients and we obtain at once the two infinite series derived in Example 1.

The method illustrated in Example 2 is essentially the standard technique by which we may obtain solutions of the equations which define all the "special functions" of mathematical physics. The equation which relates the coefficients in the series solutions is called a *recurrence relation*. Modifications of this approach are frequently necessary, mainly for two reasons: (i) the recurrence relation may connect several coefficients and be too unwieldy to use, or (ii) it may be desirable to expand about a point which is *not* an ordinary point. We now examine briefly some of the equations and eigenfunction sets of special importance in quantum mechanics, illustrating further points as they arise.

(i) *Hermite's equation*

The harmonic oscillator equation (p. 27) is of the form (putting $\alpha = 1$)

$$\frac{d^2y}{dx^2} + (\lambda - x^2)y = 0 \qquad (A3.4)$$

and provides an example in which the simple series solution leads to a recurrence relation connecting *three* consecutive coefficients. In such cases it is often possible to obtain a simpler equation by the substitution

$$y(x) = f(x)u(x) \qquad (A3.5)$$

where $f(x)$ is to be chosen so as to yield a simpler equation for $u(x)$. Two common choices are (a) to put $f(x) = x^\nu$ (ν to be determined), and (b) to take for $f(x)$ the asymptotic solution in cases where the equation is soluble for $x \to \infty$. The second substitution was used in Section 2.2 and yielded

Hermite's equation:

$$\frac{d^2u}{dx^2} - 2x\,\frac{du}{dx} + 2\alpha u = 0 \qquad (A3.6)$$

where we have introduced $2\alpha = (\lambda - 1)$.

The series solution of Hermite's equation is straightforward. The point $x = 0$ is ordinary, so we look for a solution

$$u = \sum_{m=0}^{\infty} a_m x^m$$

and obtain on substitution

$$\sum_{m=0}^{\infty} a_m m(m-1)x^{m-2} - 2\sum_{m=0}^{\infty} a_m m x^m + 2\alpha \sum_{m=0}^{\infty} a_m x^m = 0. \qquad (A3.7)$$

Terms in x^p are obtained from the first sum when $m = p+2$, and from the second and third sums when $m = p$: together they give a contribution

$$[a_{p+2}(p+2)(p+1) - 2a_p p - 2\alpha a_p]x^p$$

and on equating to zero the coefficient of each power of x we obtain the *two*-term recurrence relation

$$a_{p+2} = \frac{2(p-\alpha)}{(p+1)(p-1)}\,a_p \quad (p > 0). \qquad (A3.8)$$

Reference to (A3.7) shows, however, that there are terms (from the first sum only) in x^{-2} and x^{-1}, which vanish because of the factor $m(m-1)$

without fixing a_0 and a_1. We therefore obtain two distinct series solutions, their sum being the general solution (two arbitrary constants)

$$u = a_0\left[1 - \frac{2\alpha x^2}{2!} + \frac{2^2\alpha(\alpha-2)\,x^4}{4!} - \frac{2^3\alpha(\alpha-2)\,(\alpha-4)x^6}{6!} + \cdots\right]$$

$$+ a_1\left[x - \frac{2(\alpha-1)x^3}{3!} + \frac{2^2(\alpha-1)\,(\alpha-3)x^5}{5!} - \cdots\right] \qquad (A3.9)$$

In Section 2.2 it was stated that the solutions became infinite for $x \to \pm\infty$ more rapidly than the asymptotic solution $f(x) = e^{-\frac{1}{2}x^2}$ except for certain special choices of the parameter $(\lambda = 1+2\alpha)$: the acceptable solutions $(y = fu)$ of (A3.4) are now seen to occur when one or other of the above series terminates, so that $u(x)$ becomes a finite polynomial. If α is an even positive integer, n say, the first series terminates and becomes a polynomial of degree n; similarly, the second series gives a polynomial of odd degree when α is an odd positive integer. These special solutions are the *Hermite polynomials*.

It is customary to "normalize" the polynomials so that the term of highest degree is $(2x)^n$. We may then use (A3.6) the other way round, (putting $\alpha = n$ and replacing p by $q-2$),

$$a_{q-2} = \frac{-q(q-1)}{2(n-q+2)}\,a_q,$$

to obtain a descending series, valid for n either even or odd. This is the standard solution

$$H_n(x) = (2x)^n - n(n-1)(2x)^{n-2} + \frac{n(n-1)\,(n-2)\,(n-3)}{2!}\,(2x)^{n-4} - \cdots$$

$$(A3.10)$$

The corresponding particular solution of the original equation (A3.4) occurs when $\lambda = 2n+1$ and is

$$y_n(x) = H_n(x)e^{-\frac{1}{2}x^2} \qquad (A3.11)$$

and $y_n(x) \to 0$ for $x \to \pm\infty$. This is the *Hermite function* of order n.

(ii) *Legendre's equation*

In the central field problem, introduced in Section 2.3 and dealt with more fully in Vol. 2, Chapter 2, we meet an equation

$$(1-x^2)\frac{d^2y}{dx^2} - 2x\frac{dy}{dx} + \lambda y = 0. \qquad (A3.12)$$

This is *Legendre's equation*. Since it originates from a change of variables, with $x = \cos\theta$, we are interested mainly in solutions defined in the interval $(-1, +1)$: singularities thus occur at the end points, for if the equation were written in the form (A3.1) the functions $P(x)$ and $Q(x)$ would not be analytic at $x = \pm 1$. Nevertheless, it is safe to develop the solution about $x = 0$ which is an ordinary point.

The procedure followed in dealing with Hermite's equation here yields a recurrence relation

$$a_{p+2} = \frac{p(p+1)-\lambda}{(p+1)(p+2)}\, a_p. \tag{A3.13}$$

As in (A3.9), a_0 and a_1 are indeterminate and, putting $-\lambda = \bar{\lambda}$, a solution is obtained in the form

$$y = a_0\left[1 + \frac{\bar{\lambda}}{2!}x + \frac{(2.3+\bar{\lambda})}{4!}\, x^2 + \ldots\right]$$
$$+ a_1\left[\frac{x + (1.2+\bar{\lambda})\, x^3}{3!} + \frac{\bar{\lambda}(3.4+\bar{\lambda})}{5!}\, x^5 + \ldots\right]. \tag{A3.14}$$

The two series represent distinct solutions (odd and even), their sum being the general solution. The recurrence relation shows that if $\lambda = l(l+1)$ where l is any positive integer then all coefficients beyond a_l will vanish and one of the series terminates: the series that terminates, when conventionally normalized, is the *Legendre polynomial* $P_l(x)$.

A descending series, valid for l odd or even, is obtained as for the Hermite polynomials. The usual normalization is then obtained on taking $a_l = (2l)!/2^l(l!)^2$ which ensures that $P_l(1) = 1$ (all l). The standard polynomial, determined in this way, is thus

$$P_l(x) = \frac{(2l)!}{2^l(l!)^2}\left[x^l - \frac{l(l-1)}{2(2-1)}\, x^{l-2} + \frac{l(l-1)(l-2)(l-3)}{2.4(2l-1)(2l-3)}x^{l-4} - \ldots\right].$$
$$\tag{A3.15}$$

This is also valid as the "first solution" of Legendre's equation even when $|x| > 1$, but further developments are not required in this book.

(iii)   *Legendre's associated equation*

The more general equation (Chap. 2 of Vol. 2), of which (A3.12) is a special case, has the form

$$(1-x^2)\frac{d^2y}{dx^2} - 2x\frac{dy}{dx} + \left[l(l+1) - \frac{m^2}{1-x^2}\right]y = 0 \tag{A3.16}$$

where the parameter λ has been given its value $l(l+1)$ with l a positive integer. This is the *associated Legendre equation* and our main concern is with solutions in which m is also integral. The equation can be solved directly by the series method, but it is easier to find a relationship with the Legendre equation itself—which corresponds to $m = 0$. Consideration of the solution near the singularities at $x = \pm 1$ suggests the substitution

$$y = (1-x^2)^{m/2}u \qquad (A3.17)$$

and this leads to an equation for u:

$$(1-x^2)\frac{d^2u}{dx^2} - 2(m+1)x\frac{du}{dx} + (l-m)(l+m+1)u = 0. \qquad (A3.18)$$

This resembles the equation satisfied by $P_l(x)$, namely

$$(1-x^2)\frac{d^2P_l}{dx^2} - 2x\frac{dP_l}{dx}\ l(l+1)P_l = 0.$$

If we differentiate this latter equation m times and collect the terms we obtain, in fact

$$(1-x^2)\frac{d^2}{dx^2}\left(\frac{d^mP_l}{dx^m}\right) - 2(m+1)x\left(\frac{d^mP_l}{dx^m}\right) + (l-m)(l+m+1)\left(\frac{d^mP_l}{dx^m}\right) = 0$$

and therefore $u = d^mP_l/dx^m$ is a solution of (A3.18). From (A3.17) it follows that the associated Legendre equation has a solution

$$P_l^m(x) = (1-x^2)^{m/2}\frac{d^mP_l(x)}{dx^m}. \qquad (A3.19)$$

The result is, of course, for l and m positive integers, a polynomial of degree l which may be obtained from (A3.15) if desired.

(iv) *Laguerre's equation*

The differential equation for the radial factor in the hydrogen atom wave functions (Chap. 2 of Vol. 2) is obtained by finding first the asymptotic solution for $r \to \infty$ and then making a substitution of the form (A3.5). The reduced equation takes its simplest form for spherically symmetrical functions, namely,

$$x\frac{d^2y}{dx^2} + (1-x)\frac{dy}{dx} + \alpha y = 0. \qquad (A3.20)$$

This is *Laguerre's equation* and the solutions of interest in quantum mechanics are $L_\alpha(x)$ where α is a positive integer: they are the *Laguerre polynomials*.

If the equation (A3.20) were written in the standard form (A3.1) the function $P(x)$ would be singular at $x = 0$ and the series method of solution as used so far would break down, (d^2y/dx^2) being indeterminate at this point. It is possible to obtain a solution, however, by a simple extension of the method. In general, if an equation can be written in the form (cf. A3.1)

$$(x-a)^2 \frac{d^2y}{dx^2} + (x-a)P(x)\frac{dy}{dx} + Q(x)y = 0 \qquad (A3.21)$$

it is said to possess a *regular* singularity at $x = a$, which is called a *regular point*. It can then be shown that there exists a solution in the form

$$y = a_0(x-a)^l + a_1(x-a)^{l+1} + \ldots a_r(x-a)^{l+r} + \ldots \qquad (A3.22)$$

where the value of l is determinate on substituting this series and equating coefficients of all powers to zero in the usual way: the equation determining l is called the *indicial equation*.

The Laguerre equation is seen to have a regular singularity at $x = 0$: accordingly we look for a solution

$$y = x^l(a_0 + a_1x + a_2x^2 + \ldots) \qquad (A3.23)$$

and readily obtain, on equating to zero the coefficient of the lowest power of x, the indicial equation $l^2 = 0$. The simple series, starting with the constant term, is therefore valid in this case in spite of the singularity. The recurrence relation connecting the coefficients is

$$a_{p+1} = \frac{p-\alpha}{(p+1)^2}a_p \qquad (A3.24)$$

and this leads at once to the series solution. The series terminates when $\alpha = n$, a positive integer, and the Laguerre polynomial is defined so that the coefficient of x^n is $(-1)^n$. The result, as a descending series, is

$$L_n(x) = (-1)^n\left[x^n - \frac{n^2}{1!}x^{n-1} + \frac{n^2(n-1)^2}{2!}x^{n-2} - \ldots\right] \qquad (A3.25)$$

which is the polynomial solution of (A3.20) for the case $\alpha = n$. Angle-independent solutions of (2.20) take the form $e^{-r/n}L_n(2r/n)$.

(v) *Laguerre's associated equation*

In determining the general hydrogen-atom wave functions, there appears a generalization of (A3.20), namely

$$x\frac{d^2y}{dx^2} + (1+\beta-x)\frac{dy}{dx} + (\alpha-\beta)y = 0 \qquad (A3.26)$$

which is known as the *associated Laguerre equation*. This equation is related to (A3.20), when $\alpha(= n)$ and $\beta(= m)$ are positive integers with $m < n$, as may be verified by differentiating the latter m times: it is then apparent that the $(n - m)$th degree polynomial

$$L_n{}^m(x) = \frac{d^m}{dx^m} L_n(x) \qquad (A3.27)$$

is a solution of (A3.26) with $\alpha = n$, $\beta = m$. $L_n{}^m(x)$ is the *associated Laguerre polynomial*.

A more detailed discussion of the functions of Hermite, Legendre and Laguerre may be found in the book by Sneddon (1956).

REFERENCE

SNEDDON, I. N. *The Special Functions of Mathematical Physics and Chemistry*, Oliver & Boyd, Edinburgh and London.

APPENDIX 4

PROJECTION OPERATORS AND NORMAL FORMS

In ELEMENTARY vector algebra it is sometimes convenient (e.g. Morse and Feshbach, 1953) to introduce "dyads" which provide a symbolic representation of certain operators. Thus $e_i e_j$, in which no scalar product is implied, denotes the operator which when applied to a vector r has the effect

$$(e_i e_j)r = e_i(e_j \cdot r),$$

i.e. the right-hand factor of the dyad is combined with the operand to form a scalar product: the result is thus a scalar multiple of the first vector in the dyad. The dyads $e_i e_i$ are particularly important because they describe *projection operators*. Thus, assuming as usual an orthonormal basis, we obtain from an arbitrary vector $r = r_i e_i + r_2 e_2 + r_3 e_3$

$$(e_1 e_1)r = e_1[(r_1 e_1 \cdot e_1) + (r_2 e_1 \cdot e_2) + (r_3 e_1 \cdot e_3)] = r_1 e_1. \quad (A4.1)$$

The operator $(e_1 e_1)$ annihilates all components of r except that in the direction e_1, which it leaves unchanged: thus it produces the projection of r along the e_1 axis.

In a Hermitian vector space similar considerations apply, except that $e_i e_j$ is replaced by $e_i e_j * \ddagger$ or in a function space by $\Phi_i \Phi_j *$. Here we use the symbolic notation of an earlier section (p. 57). Thus if

$$\Psi = c_1 \Phi_1 + c_2 \Phi_2 + \ldots + c_k \Phi_k + \ldots$$

is an arbitrary element of the space, and we assume orthonormality of the set $\{\Phi_i\}$, the operator $\Phi_i \Phi_i *$ is interpreted as in (A4.1):

$$(\Phi_i \Phi_i *)\Psi = \Phi_i[c_1(\Phi_i * \Phi_1) + c_2(\Phi_i * \Phi_2) + \ldots c_k(\Phi_i * \Phi_k) + \ldots] = c_i \Phi_i$$

$$(A4.2)$$

—since every scalar product $\Phi_i * \Phi_k = \langle \Phi_i | \Phi_k \rangle$ vanishes except that with $k = i$. Consequently $(\Phi_i \Phi_i *)$, interpreted in this way, picks out the ith component of an arbitrary vector Ψ.

Now let us consider a general operator R formed as a linear combination of dyads $\Phi_i \Phi_j *$ with numerical coefficients:

$$R = \sum_{k,l} M_{kl}(\Phi_k \Phi_l *). \quad (A4.3)$$

---

‡Note that the star must be on the right, otherwise a scalar product would be implied.

With R we may associate a matrix in the usual way, its elements being determined by

$$R_{ij} = \langle \Phi_i | R | \Phi_j \rangle$$
$$= \Phi_i^* \sum_{k,l} M_{kl} (\Phi_k \Phi_l^*) \Phi_j$$
$$= \sum_{k,l} M_{kl} (\Phi_i^* \Phi_k)(\Phi_l^* \Phi_j)$$
$$= M_{ij}.$$

Thus any operator R may be expressed in "dyadic form", the coefficients in (A4.3) being its matrix elements in the usual sense:

$$R = \sum_{ij} R_{ij} \Phi_i \Phi_j^*. \tag{A4.4}$$

It should be noted that such considerations apply with an *orthonormal basis* but need modification otherwise.

Suppose now we take an operator such as H in (3.70) and choose its eigenfunctions as the basis vectors of the representation space. Then, from (3.75), it follows that the dyadic form of H is

$$H = \sum_K E_K (\Psi_K \Psi_K^*) = \sum_k E_K \mathsf{P}_K \tag{A4.5}$$

where we have used P_K to denote the projection operator onto the Kth basis function. This representation of an operator, in terms of its eigenvalues and the projection operators onto its eigenvectors, is referred to as a "normal form"; this form plays an important part in the rigorous formulation of quantum mechanics, particularly in the extension to operators with a continuous spectrum (von Neumann, 1955).

The basic properties of projection operators are

(a) $\mathsf{P}_K{}^2 = \mathsf{P}_K$

(b) $\mathsf{P}_K \mathsf{P}_L = \mathsf{P}_L \mathsf{P}_K = 0 \quad (K \neq L),$ (A4.6)

(c) $\sum_K \mathsf{P}_K = 1$

where (a) is described as "idempotency". In (b) and (c), 0 and 1 are interpreted as zero and unit *operators* (i.e. annihilating a function or leaving it unchanged). Thus (b) means that successive projection of a vector onto two orthogonal axes leaves nothing. On the other hand, (c) means that projection onto *all* axes, followed by recombination, restores the original vector—a result referred to as "resolution of the identity".

An immediate application of the projection operators is to the definition of a *function* of an operator: thus

$$\mathsf{H}^2 = \Big(\sum_K E_K \mathsf{P}_K\Big)\Big(\sum_L E_L \mathsf{P}_L\Big) = \sum_K E_K{}^2 \mathsf{P}_K,$$

which clearly extends to all powers of H, suggests the definition

$$f(\mathsf{H}) = \sum_K f(E_K)\mathsf{P}_K. \tag{A4.7}$$

$f(\mathsf{H})$ is thus the operator whose eigenvalues are the same function of the eigenvalues of H.

It is not difficult to translate the above considerations into matrix form, using an arbitrary orthonormal basis. With the eigenvectors Ψ_1, Ψ_2, ... Ψ_K, ... we then associate columns $\mathbf{c}_1$, $\mathbf{c}_2$, ... $\mathbf{c}_K$, ... while with the projection operator P_K we associate the square matrix (column–row product)

$$\mathbf{P}_K = \mathbf{c}_K \mathbf{c}_K{}^\dagger \tag{A4.8}$$

whose ij-element is $\mathbf{P}_{Kij} = c_{Ki} c_{Kj}{}^*$. The effect of $\mathbf{P}_K$ on a column $\mathbf{c}$, representing an arbitrary vector, is to yield a multiple of $\mathbf{c}_K$. The *matrix* $\mathbf{H}$, associated with H, is then expressed in a normal form resembling (A4.5):

$$\mathbf{H} = \sum_K E_K \mathbf{c}_K \mathbf{c}_K{}^\dagger = \sum_K E_K \mathbf{P}_K. \tag{A4.9}$$

As usual, there is a complete parallel between the operator and the matrix forms, extending for example to the properties (A4.6) provided that 0 and 1 are interpreted as zero and unit matrices.

Projection onto a *many*-dimensional subspace may be defined analogously in terms of a sum: thus $\mathsf{P} = \mathsf{P}_1 + \mathsf{P}_2 + \ldots \mathsf{P}_L$ annihilates all components of an arbitrary vector except those referring to the first L basis functions Ψ_1, Ψ_2, ... Ψ_L. If we use $\mathbf{T}$ to denote the matrix whose L columns are $\mathbf{c}_1$, $\mathbf{c}_2$, ... $\mathbf{c}_L$, it is clear that

$$\mathbf{P} = \mathbf{c}_1 \mathbf{c}_1{}^\dagger + \mathbf{c}_2 \mathbf{c}_2{}^\dagger + \ldots \mathbf{c}_L \mathbf{c}_L{}^\dagger = \mathbf{T}\mathbf{T}^\dagger. \tag{A4.10}$$

This matrix also possesses the idempotency property $\mathbf{P}^2 = \mathbf{P}$ (cf. A4.6a). The remaining functions $\{\Psi_J (J > L)\}$ also define a subspace, the *orthogonal complement* of that defined by the projection operator P. It has a complementary projection operator $(1-\mathsf{P})$ and an arbitrary function may then be resolved into parts lying inside and outside a given subspace:

$$\Psi = \mathsf{P}\Psi + (1-\mathsf{P})\Psi = \Psi' + \Psi''. \tag{A4.11}$$

Clearly P and $1-\mathsf{P}$ are orthogonal operators in the sense $\mathsf{P}(1-\mathsf{P}) = \mathsf{P} - \mathsf{P}^2 = 0$ (cf. (A4.6b)), whilst their sum provides a resolution of the identity (cf. (A4.6c)).

The use of projection operators gives a convenient method of discussing the effects of truncation in complete set expansions. Again it must be stressed that the complete equivalence of operator and matrix equations applies only in the limit of completeness ($n \to \infty$) and is subject to convergence considerations. In any finite formulation the matrix procedures may be used freely: but any conclusions then refer not to the original operator R (infinite summations in (A4.4)) but to its projection $\hat{R}$ on the given subspace (defined by truncating the summations in (A4.11)). This simply means, for example, that the eigenvalues and eigenvectors arising from a *finite* matrix representation of the operator equation $H\Psi = E\Psi$ are not solutions of the actual Schrödinger equation but rather of its "projection within a subspace".

REFERENCES

MORSE, P. M. and FESHBACH, H. (1953) *Methods of Theoretical Physics,* Vol. I, McGraw Hill, New York. (See, in particular, Section 1.6.)

VON NEUMANN, J. (1955) *Mathematical Foundations of Quantum Mechanics,* Princeton University Press, Princeton. (Translated from the German edition by R. T. Beyer.)

INDEX

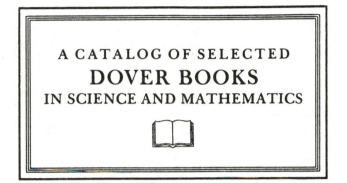

A CATALOG OF SELECTED
DOVER BOOKS
IN SCIENCE AND MATHEMATICS

A CATALOG OF SELECTED
DOVER BOOKS
IN SCIENCE AND MATHEMATICS

Astronomy

BURNHAM'S CELESTIAL HANDBOOK, Robert Burnham, Jr. Thorough guide to the stars beyond our solar system. Exhaustive treatment. Alphabetical by constellation: Andromeda to Cetus in Vol. 1; Chamaeleon to Orion in Vol. 2; and Pavo to Vulpecula in Vol. 3. Hundreds of illustrations. Index in Vol. 3. 2,000pp. 6⅛ x 9¼.
23567-X, 23568-8, 23673-0 Three-vol. set

THE EXTRATERRESTRIAL LIFE DEBATE, 1750–1900, Michael J. Crowe. First detailed, scholarly study in English of the many ideas that developed from 1750 to 1900 regarding the existence of intelligent extraterrestrial life. Examines ideas of Kant, Herschel, Voltaire, Percival Lowell, many other scientists and thinkers. 16 illustrations. 704pp. 5⅜ x 8½.
40675-X

A HISTORY OF ASTRONOMY, A. Pannekoek. Well-balanced, carefully reasoned study covers such topics as Ptolemaic theory, work of Copernicus, Kepler, Newton, Eddington's work on stars, much more. Illustrated. References. 521pp. 5⅜ x 8½.
65994-1

AMATEUR ASTRONOMER'S HANDBOOK, J. B. Sidgwick. Timeless, comprehensive coverage of telescopes, mirrors, lenses, mountings, telescope drives, micrometers, spectroscopes, more. 189 illustrations. 576pp. 5⅜ x 8¼. (Available in U.S. only.)
24034-7

STARS AND RELATIVITY, Ya. B. Zel'dovich and I. D. Novikov. Vol. 1 of *Relativistic Astrophysics* by famed Russian scientists. General relativity, properties of matter under astrophysical conditions, stars, and stellar systems. Deep physical insights, clear presentation. 1971 edition. References. 544pp. 5⅜ x 8¼.   69424-0

Chemistry

CHEMICAL MAGIC, Leonard A. Ford. Second Edition, Revised by E. Winston Grundmeier. Over 100 unusual stunts demonstrating cold fire, dust explosions, much more. Text explains scientific principles and stresses safety precautions. 128pp. 5⅜ x 8½.
67628-5

THE DEVELOPMENT OF MODERN CHEMISTRY, Aaron J. Ihde. Authoritative history of chemistry from ancient Greek theory to 20th-century innovation. Covers major chemists and their discoveries. 209 illustrations. 14 tables. Bibliographies. Indices. Appendices. 851pp. 5⅜ x 8½.
64235-6

CATALYSIS IN CHEMISTRY AND ENZYMOLOGY, William P. Jencks. Exceptionally clear coverage of mechanisms for catalysis, forces in aqueous solution, carbonyl- and acyl-group reactions, practical kinetics, more. 864pp. 5⅜ x 8½.
65460-5

Math–Geometry and Topology

ELEMENTARY CONCEPTS OF TOPOLOGY, Paul Alexandroff. Elegant, intuitive approach to topology from set-theoretic topology to Betti groups; how concepts of topology are useful in math and physics. 25 figures. 57pp. 5⅜ x 8½. 60747-X

COMBINATORIAL TOPOLOGY, P. S. Alexandrov. Clearly written, well-organized, three-part text begins by dealing with certain classic problems without using the formal techniques of homology theory and advances to the central concept, the Betti groups. Numerous detailed examples. 654pp. 5⅜ x 8½. 40179-0

EXPERIMENTS IN TOPOLOGY, Stephen Barr. Classic, lively explanation of one of the byways of mathematics. Klein bottles, Moebius strips, projective planes, map coloring, problem of the Koenigsberg bridges, much more, described with clarity and wit. 43 figures. 210pp. 5⅜ x 8½. 25933-1

CONFORMAL MAPPING ON RIEMANN SURFACES, Harvey Cohn. Lucid, insightful book presents ideal coverage of subject. 334 exercises make book perfect for self-study. 55 figures. 352pp. 5⅜ x 8¼. 64025-6

THE GEOMETRY OF RENÉ DESCARTES, René Descartes. The great work founded analytical geometry. Original French text, Descartes's own diagrams, together with definitive Smith-Latham translation. 244pp. 5⅜ x 8½. 60068-8

THE THIRTEEN BOOKS OF EUCLID'S ELEMENTS, translated with introduction and commentary by Sir Thomas L. Heath. Definitive edition. Textual and linguistic notes, mathematical analysis. 2,500 years of critical commentary. Unabridged. 1,414pp. 5⅜ x 8½. Three-vol. set.
Vol. I: 60088-2   Vol. II: 60089-0   Vol. III: 60090-4

GEOMETRY OF COMPLEX NUMBERS, Hans Schwerdtfeger. Illuminating, widely praised book on analytic geometry of circles, the Moebius transformation, and two-dimensional non-Euclidean geometries. 200pp. 5⅜ x 8¼. 63830-8

DIFFERENTIAL GEOMETRY, Heinrich W. Guggenheimer. Local differential geometry as an application of advanced calculus and linear algebra. Curvature, transformation groups, surfaces, more. Exercises. 62 figures. 378pp. 5⅜ x 8½. 63433-7

CURVATURE AND HOMOLOGY: Enlarged Edition, Samuel I. Goldberg. Revised edition examines topology of differentiable manifolds; curvature, homology of Riemannian manifolds; compact Lie groups; complex manifolds; curvature, homology of Kaehler manifolds. New Preface. Four new appendixes. 416pp. 5⅜ x 8½. 40207-X

TOPOLOGY, John G. Hocking and Gail S. Young. Superb one-year course in classical topology. Topological spaces and functions, point-set topology, much more. Examples and problems. Bibliography. Index. 384pp. 5⅜ x 8¼. 65676-4

LECTURES ON CLASSICAL DIFFERENTIAL GEOMETRY, Second Edition, Dirk J. Struik. Excellent brief introduction covers curves, theory of surfaces, fundamental equations, geometry on a surface, conformal mapping, other topics. Problems. 240pp. 5⅜ x 8½. 65609-8

Math-History of

A SHORT ACCOUNT OF THE HISTORY OF MATHEMATICS, W. W. Rouse Ball. One of clearest, most authoritative surveys from the Egyptians and Phoenicians through 19th-century figures such as Grassman, Galois, Riemann. Fourth edition. 522pp. 5⅜ x 8½. 20630-0

THE HISTORY OF THE CALCULUS AND ITS CONCEPTUAL DEVELOPMENT, Carl B. Boyer. Origins in antiquity, medieval contributions, work of Newton, Leibniz, rigorous formulation. Treatment is verbal. 346pp. 5⅜ x 8½. 60509-4

THE HISTORICAL ROOTS OF ELEMENTARY MATHEMATICS, Lucas N. H. Bunt, Phillip S. Jones, and Jack D. Bedient. Fundamental underpinnings of modern arithmetic, algebra, geometry and number systems derived from ancient civilizations. 320pp. 5⅜ x 8½. 25563-8

A HISTORY OF MATHEMATICAL NOTATIONS, Florian Cajori. This classic study notes the first appearance of a mathematical symbol and its origin, the competition it encountered, its spread among writers in different countries, its rise to popularity, its eventual decline or ultimate survival. Original 1929 two-volume edition presented here in one volume. xxviii+820pp. 5⅜ x 8½. 67766-4

GAMES, GODS & GAMBLING: A History of Probability and Statistical Ideas, F. N. David. Episodes from the lives of Galileo, Fermat, Pascal, and others illustrate this fascinating account of the roots of mathematics. Features thought-provoking references to classics, archaeology, biography, poetry. 1962 edition. 304pp. 5⅜ x 8½. (Available in U.S. only.) 40023-9

OF MEN AND NUMBERS: The Story of the Great Mathematicians, Jane Muir. Fascinating accounts of the lives and accomplishments of history's greatest mathematical minds—Pythagoras, Descartes, Euler, Pascal, Cantor, many more. Anecdotal, illuminating. 30 diagrams. Bibliography. 256pp. 5⅜ x 8½. 28973-7

HISTORY OF MATHEMATICS, David E. Smith. Nontechnical survey from ancient Greece and Orient to late 19th century; evolution of arithmetic, geometry, trigonometry, calculating devices, algebra, the calculus. 362 illustrations. 1,355pp. 5⅜ x 8½. Two-vol. set. Vol. I: 20429-4 Vol. II: 20430-8

A CONCISE HISTORY OF MATHEMATICS, Dirk J. Struik. The best brief history of mathematics. Stresses origins and covers every major figure from ancient Near East to 19th century. 41 illustrations. 195pp. 5⅜ x 8½. 60255-9

Physics

OPTICAL RESONANCE AND TWO-LEVEL ATOMS, L. Allen and J. H. Eberly. Clear, comprehensive introduction to basic principles behind all quantum optical resonance phenomena. 53 illustrations. Preface. Index. 256pp. 5⅜ x 8½.   65533-4

ULTRASONIC ABSORPTION: An Introduction to the Theory of Sound Absorption and Dispersion in Gases, Liquids and Solids, A. B. Bhatia. Standard reference in the field provides a clear, systematically organized introductory review of fundamental concepts for advanced graduate students, research workers. Numerous diagrams. Bibliography. 440pp. 5⅜ x 8½.   64917-2

QUANTUM THEORY, David Bohm. This advanced undergraduate-level text presents the quantum theory in terms of qualitative and imaginative concepts, followed by specific applications worked out in mathematical detail. Preface. Index. 655pp. 5⅜ x 8½.   65969-0

ATOMIC PHYSICS (8th edition), Max Born. Nobel laureate's lucid treatment of kinetic theory of gases, elementary particles, nuclear atom, wave-corpuscles, atomic structure and spectral lines, much more. Over 40 appendices, bibliography. 495pp. 5⅜ x 8½.   65984-4

AN INTRODUCTION TO HAMILTONIAN OPTICS, H. A. Buchdahl. Detailed account of the Hamiltonian treatment of aberration theory in geometrical optics. Many classes of optical systems defined in terms of the symmetries they possess. Problems with detailed solutions. 1970 edition. xv + 360pp. 5⅜ x 8½.   67597-1

THIRTY YEARS THAT SHOOK PHYSICS: The Story of Quantum Theory, George Gamow. Lucid, accessible introduction to influential theory of energy and matter. Careful explanations of Dirac's anti-particles, Bohr's model of the atom, much more. 12 plates. Numerous drawings. 240pp. 5⅜ x 8½.   24895-X

ELECTRONIC STRUCTURE AND THE PROPERTIES OF SOLIDS: The Physics of the Chemical Bond, Walter A. Harrison. Innovative text offers basic understanding of the electronic structure of covalent and ionic solids, simple metals, transition metals and their compounds. Problems. 1980 edition. 582pp. 6⅛ x 9¼.   66021-4

HYDRODYNAMIC AND HYDROMAGNETIC STABILITY, S. Chandrasekhar. Lucid examination of the Rayleigh-Benard problem; clear coverage of the theory of instabilities causing convection. 704pp. 5⅜ x 8¼.   64071-X

INVESTIGATIONS ON THE THEORY OF THE BROWNIAN MOVEMENT, Albert Einstein. Five papers (1905–8) investigating dynamics of Brownian motion and evolving elementary theory. Notes by R. Fürth. 122pp. 5⅜ x 8½.   60304-0

THE PHYSICS OF WAVES, William C. Elmore and Mark A. Heald. Unique overview of classical wave theory. Acoustics, optics, electromagnetic radiation, more. Ideal as classroom text or for self-study. Problems. 477pp. 5⅜ x 8½.   64926-1

CATALOG OF DOVER BOOKS

PHYSICAL PRINCIPLES OF THE QUANTUM THEORY, Werner Heisenberg. Nobel Laureate discusses quantum theory, uncertainty, wave mechanics, work of Dirac, Schroedinger, Compton, Wilson, Einstein, etc. 184pp. 5⅜ x 8½.   60113-7

ATOMIC SPECTRA AND ATOMIC STRUCTURE, Gerhard Herzberg. One of best introductions; especially for specialist in other fields. Treatment is physical rather than mathematical. 80 illustrations. 257pp. 5⅜ x 8½.   60115-3

AN INTRODUCTION TO STATISTICAL THERMODYNAMICS, Terrell L. Hill. Excellent basic text offers wide-ranging coverage of quantum statistical mechanics, systems of interacting molecules, quantum statistics, more. 523pp. 5⅜ x 8½.   65242-4

THEORETICAL PHYSICS, Georg Joos, with Ira M. Freeman. Classic overview covers essential math, mechanics, electromagnetic theory, thermodynamics, quantum mechanics, nuclear physics, other topics. First paperback edition. xxiii + 885pp. 5⅜ x 8½.   65227-0

PROBLEMS AND SOLUTIONS IN QUANTUM CHEMISTRY AND PHYSICS, Charles S. Johnson, Jr. and Lee G. Pedersen. Unusually varied problems, detailed solutions in coverage of quantum mechanics, wave mechanics, angular momentum, molecular spectroscopy, more. 280 problems plus 139 supplementary exercises. 430pp. 6½ x 9¼.   65236-X

THEORETICAL SOLID STATE PHYSICS, Vol. 1: Perfect Lattices in Equilibrium; Vol. II: Non-Equilibrium and Disorder, William Jones and Norman H. March. Monumental reference work covers fundamental theory of equilibrium properties of perfect crystalline solids, non-equilibrium properties, defects and disordered systems. Appendices. Problems. Preface. Diagrams. Index. Bibliography. Total of 1,301pp. 5⅜ x 8½. Two volumes.   Vol. I: 65015-4   Vol. II: 65016-2

A TREATISE ON ELECTRICITY AND MAGNETISM, James Clerk Maxwell. Important foundation work of modern physics. Brings to final form Maxwell's theory of electromagnetism and rigorously derives his general equations of field theory. 1,084pp. 5⅜ x 8½. Two-vol. set.   Vol. I: 60636-8   Vol. II: 60637-6

OPTICKS, Sir Isaac Newton. Newton's own experiments with spectroscopy, colors, lenses, reflection, refraction, etc., in language the layman can follow. Foreword by Albert Einstein. 532pp. 5⅜ x 8½.   60205-2

THEORY OF ELECTROMAGNETIC WAVE PROPAGATION, Charles Herach Papas. Graduate-level study discusses the Maxwell field equations, radiation from wire antennas, the Doppler effect and more. xiii + 244pp. 5⅜ x 8½.   65678-5

INTRODUCTION TO QUANTUM MECHANICS With Applications to Chemistry, Linus Pauling & E. Bright Wilson, Jr. Classic undergraduate text by Nobel Prize winner applies quantum mechanics to chemical and physical problems. Numerous tables and figures enhance the text. Chapter bibliographies. Appendices. Index. 468pp. 5⅜ x 8½.   64871-0

CATALOG OF DOVER BOOKS

METHODS OF THERMODYNAMICS, Howard Reiss. Outstanding text focuses on physical technique of thermodynamics, typical problem areas of understanding, and significance and use of thermodynamic potential. 1965 edition. 238pp. 5⅜ x 8½.
69445-3

TENSOR ANALYSIS FOR PHYSICISTS, J. A. Schouten. Concise exposition of the mathematical basis of tensor analysis, integrated with well-chosen physical examples of the theory. Exercises. Index. Bibliography. 289pp. 5⅜ x 8½.
65582-2

RELATIVITY IN ILLUSTRATIONS, Jacob T. Schwartz. Clear nontechnical treatment makes relativity more accessible than ever before. Over 60 drawings illustrate concepts more clearly than text alone. Only high school geometry needed. Bibliography. 128pp. 6⅛ x 9¼.
25965-X

THE ELECTROMAGNETIC FIELD, Albert Shadowitz. Comprehensive undergraduate text covers basics of electric and magnetic fields, builds up to electromagnetic theory. Also related topics, including relativity. Over 900 problems. 768pp. 5⅜ x 8¼.
65660-8

GREAT EXPERIMENTS IN PHYSICS: Firsthand Accounts from Galileo to Einstein, edited by Morris H. Shamos. 25 crucial discoveries: Newton's laws of motion, Chadwick's study of the neutron, Hertz on electromagnetic waves, more. Original accounts clearly annotated. 370pp. 5⅜ x 8½.
25346-5

RELATIVITY, THERMODYNAMICS AND COSMOLOGY, Richard C. Tolman. Landmark study extends thermodynamics to special, general relativity; also applications of relativistic mechanics, thermodynamics to cosmological models. 501pp. 5⅜ x 8½.
65383-8

LIGHT SCATTERING BY SMALL PARTICLES, H. C. van de Hulst. Comprehensive treatment including full range of useful approximation methods for researchers in chemistry, meteorology and astronomy. 44 illustrations. 470pp. 5⅜ x 8½.
64228-3

STATISTICAL PHYSICS, Gregory H. Wannier. Classic text combines thermodynamics, statistical mechanics and kinetic theory in one unified presentation of thermal physics. Problems with solutions. Bibliography. 532pp. 5⅜ x 8½.
65401-X

Paperbound unless otherwise indicated. Available at your book dealer, online at **www.doverpublications.com**, or by writing to Dept. GI, Dover Publications, Inc., 31 East 2nd Street, Mineola, NY 11501. For current price information or for free catalogues (please indicate field of interest), write to Dover Publications or log on to **www.doverpublications.com** and see every Dover book in print. Dover publishes more than 500 books each year on science, elementary and advanced mathematics, biology, music, art, literary history, social sciences, and other areas.

THE
CRITICAL
THINKER

Second Edition

Richard Mayer
Fiona Goodchild

University of California, Santa Barbara

Brown & Benchmark
PUBLISHERS

Madison   Dubuque, IA   Guilford, CT   Chicago   Toronto   London
Caracas   Mexico City   Buenos Aires   Madrid   Bogota   Sydney

Contents

Preface

"It is strange that we expect students to learn yet seldom teach them anything about learning." This observation by Donald Norman (1980, p. 97) helps to pinpoint the goal of *The Critical Thinker*. Our goal is to explicitly teach you strategies for understanding and evaluating material in your introductory psychology textbook.

Parts of This Book

This book is intended for students who are taking their first psychology course. If you can give us six hours of your time, we think we can help you to improve how you learn and think about psychology. Here's what you will learn in this book:

Chapter 1. Critical Thinking in Psychology
  Approximate Time Required: 15 minutes
  Purpose: To convince you to commit yourself to becoming a more critical thinker.
  Topics: Reasons for improving your critical thinking skills, six aspects of the definition of critical thinking in psychology, three changes that occur as you become a critical thinker.

Chapter 2. Arguments: The Basic Units of Critical Thinking in Psychology
  Approximate Time Required: 60 minutes
  Purpose: To help you learn how to identify and analyze psychological arguments in your psychology textbook.
  Topics: Definition of an argument in psychology, three parts of an argument in psychology, four types of assertions, four types of empirical evidence, characteristics of a theoretical explanation.

Chapter 3. Strategies for Understanding and Evaluating Arguments
  Approximate Time Required: 60 minutes
  Purpose: To introduce you to some specific strategies for how to understand and evaluate the arguments presented in your psychology textbook.

Topics: How to summarize assertions, evidence, and explanations in your own words; how to criticize assertions, evidence, and explanations in your own words.

Chapters 4-8. Critical Thinking Exercises
Approximate Time Required: 45 minutes each
Purpose: To give you practice in applying strategies for understanding and evaluating arguments in psychology textbooks.
Topics: Hands-on practice in summarizing and criticizing textbook sections in social psychology, developmental psychology, cognitive psychology, motivation, and biopsychology.

Features of This Book

Previous research has highlighted three characteristics of successful programs for teaching learning and thinking skills (Mayer, 1987, 1990). This book is designed with each of these three characteristics in mind.

What to teach: Specific skills or general ability. In learning critical thinking you acquire many specific skills rather than improve one general ability. In particular, you will learn skills for understanding and evaluating arguments. It is important that you clearly comprehend these two crucial skills.
How to teach: Process or product. In learning critical thinking you should actively engage in thinking and learning processes rather than focus only on producing the correct answer. In particular, you will engage in many "hands-on" activities and practice exercises that focus on your thinking and learning processes. It is important that you take the time to participate in each of the activities presented, including filling in the blanks for each exercise.
Where to teach: Domain-specific versus domain-free. In learning critical thinking for psychology, you should learn skills specifically adapted to reading psychology textbooks rather than broad study skills that are not focused within the context of psychology textbooks. In particular, this book provides activities and exercises that come only from psychology textbooks. It is important that you try to apply what you have learned to material in your own psychology textbook.

In summary, you will actively learn and practice specific strategies for understanding and evaluating material in psychology textbooks. The goal of this book is quite straightforward: we want you to become a more effective and critical processor of information in your psychology textbook.

About the Authors

You may be curious about the authors. Richard E. Mayer is Professor and former Chair of Psychology at the University of California, Santa Barbara, and has previously taught psychology at the University of Michigan and Indiana University. Having received his Ph.D. in Psychology from the University of Michigan, Dr. Mayer has conducted research for twenty years on how to help college students learn from science and mathematics textbooks. He has authored or coauthored twelve books, including *Educational Psychology: A Cognitive Approach,* and has produced over 100 research publications, mainly in the areas of cognitive and educational psychology.

Fiona Goodchild is an Educational Outreach Director at the University of California, Santa Barbara. Previously she served for twelve years as a Counselor in the Learning Skills Unit at the University of Western Ontario. As well as developing instructional materials for college students, she has pursued research and coordinated workshops on how to improve learning strategies for college students. Dr. Goodchild holds an Ed.D. in Applied Psychology from the University of Toronto, where she studied at the Ontario Institute for Studies in Education. She is the coauthor of *Learning for Success: Strategies for Canadian Students.*

If you have any comments about this book, including suggestions for improvements, please write to us at:

Richard E. Mayer
Department of Psychology
University of California
Santa Barbara, CA 93106

Fiona M. Goodchild
Educational Outreach Director
QUEST and Materials Research
University of California
Santa Barbara, CA 93106

Acknowledgments

We wish to thank Michael D. Lange and the staff at Brown & Benchmark Publishers for their encouragement and assistance on this project. Furthermore, we are grateful to Benjamin B. Lahey, John W. Santrock, and Lester Sdorow for allowing us to use material from their excellent psychology textbooks as examples. We also appreciate the support of our families: Richard E. Mayer thanks Beverly, Kenny, David, and Sarah for their encouragement and understanding; Fiona Goodchild thanks Michael, Penny, Katie, and Anne for their suggestions and support. Finally we acknowledge the thousands of students who each year open an introductory psychology textbook for the first time. We dedicate this book to you, who we hope will become the critical thinkers of Psychology 101.

A Note to the Instructor

What Is This Book?

Students in an introductory psychology class are often novices to the discipline. Not only do they lack knowledge of psychology, but they also have limited experience in effectively reading psychology textbooks. In short, a substantial number of your students may lack the academic study skills that they need to succeed in your course. Although students may be quite proficient at "reading every word" in the textbook, they often need training in skills that would help them become more critical thinkers in psychology. For example, when faced with reading an introductory psychology textbook, some students may have difficulty in determining what is important, in organizing the material, and in making critical judgments about the material.

This book is intended to help your students become more effective psychology students by improving their ability to understand and evaluate passages from psychology textbooks. It differs from other books on critical thinking or study skills because it is designed specifically to be used with your introductory psychology textbook.

How Can I Use This Book?

The Critical Thinker can be used in several ways.

As a stand-alone tutorial. First, this book can be used as a stand-alone adjunct to your course. You need to introduce your students to *The Critical Thinker:* instruct them to carefully read each chapter and to participate fully in each activity and exercise. Point out to your students that if they are willing to make a six-hour commitment, *The Critical Thinker* can help them become more effective psychology students. Because this book is a self-contained six-hour tutorial, no further guidance from you is required.

In discussion sections. Second, this book can be incorporated into the discussion sections of your class. We recommend covering chapters 1 through 3 in a whole-class format, with particular emphasis on the material in table 3.1 and table 3.5. You should describe your thinking process as you respond to the questions in table 3.2 and table 3.6 for a passage from your textbook.

For each of the exercises in chapters 4 through 8, we recommend that you break the class into small groups of two students each. Within the group, one student should be assigned the job of understanding and evaluating a textbook passage by writing summaries using the six-part format in table 3.1 and table 3.5. The other student should be charged with identifying problems or deficiencies in what the other student has written for each of the six parts. Together, the two students should produce a final version that satisfies both of them.

Following this small group exercise, we recommend returning to the whole-class format where each group reports its results to the class; you should keep a running summary on the blackboard or overhead transparency of what each group produced for each of the six major parts of the assignment. In class discussion, you should help the class to identify the processes that were used in understanding and evaluating arguments in psychology textbooks.

Then, for the next exercise, we recommend that you reverse the roles of the students in each group so that students get practice both in writing and critiquing summaries. If time permits, the most efficient procedure would be to assign two passages to each group, with one student assigned the job of writing the summary for the first passage and critiquing the summary of the second passage, and the other student in charge of critiquing the summary of the first passage and writing the summary of the second passage.

In a workshop. Third, you can use this book as part of a minicourse or workshop on study skills for students who have been having academic difficulties in psychology. Research on expertise has encouraged the idea that students need to work on examples from a specific context or domain when they are acquiring learning and thinking skills. The materials in this book can be used in a series of sessions ranging from one to several hours in length, and are appropriate both for individual counseling and group instruction.

We welcome your comments and suggestions, and we hope your students enjoy using *The Critical Thinker* as much as we have enjoyed writing it.

Part 1

Discovering Strategies for Critical Thinking

Chapter **1**   **Critical Thinking in Psychology**

Why Should You Read This Book?

"We should be teaching students how to think. Instead, we are teaching them what to think." This quote, taken from Clement & Lochhead's (1980) *Cognitive Process Instruction,* summarizes the rationale for this book that you are holding in your hands. We want to help you to become a critical thinker; particularly, we want to help you to learn how to think critically about psychological ideas.

Most courses in psychology serve two broad goals: (1) to help you acquire some basic knowledge about psychology, and (2) to help you learn to think like a psychologist. The first goal—focusing on "what to think"—is often so overwhelmingly obvious that students may devote all their efforts only to acquiring basic knowledge. In other words, some students see their only task in a psychology course as to memorize as many facts as possible. For example, in reading a section on *positive reinforcement,* you may memorize the definition of each key term that is presented.

The second goal—focusing on "how to think"—is sometimes so subtle that the students may fail to recognize it. Yet, instructors expect you to be able to think critically about psychological ideas and concepts. For example, in reading a section on *positive reinforcement,* you may ask yourself, "What claims are being made about the power of positive reinforcement?" and "Are these claims justified?"

The theme of this book is that for you to become a successful psychology student, you need both basic knowledge and critical thinking strategies for using that knowledge. This book is intended to help you in your attempts to become a critical thinker about psychology.

What Is Critical Thinking in Psychology?

Although teachers may agree that critical thinking is a fundamental component in college-level psychology courses, they do not agree on exactly how to define critical thinking (McPeck, 1981). We define critical thinking as an active and systematic attempt to understand and evaluate arguments. This definition has six parts:

Critical thinking is an active process. When critical thinkers read their psychology textbooks, they do more than simply read every word and look at every illustration. As a critical thinker, you need to search for meaning in what you read, that is, to ask yourself, "Does this make sense?"

Critical thinking is a systematic process. In searching for meaning, critical thinkers use logical techniques. As a critical thinker, you need to go about the process of making sense in a step-by-step manner. In short, you need to ask the question, "How should I go about understanding what the author is saying?"

Critical thinking is based on arguments. The basic unit of analysis in critical thinking is the argument. An argument begins with an assertion about the properties of some object (e.g., short-term memory has a limited capacity) or the relationship between two or more objects (e.g., the more motivated a person is, the better he or she will perform on any task); and provides evidence (usually from research studies) to support and/or refute the assertion. You need to be able to recognize and analyze arguments into their parts.

Critical thinking involves understanding arguments. A critical thinker is able to recognize arguments that the author makes in a psychology textbook. As a critical thinker you need to be able to identify the parts of the argument (such as the assertion and supporting evidence) and to state the argument in your own words. In short, you need to able to answer the question, "What argument is the author making?"

Critical thinking involves evaluating arguments. A critical thinker not only understands the author's arguments, but can also criticize them. In becoming a critical thinker, you need to learn how to determine whether the author's argument is valid; that is, you need to ask, "Should I accept the author's argument?"

Critical thinking is an attempt. Critical thinkers recognize that there is no single right way to understand and evaluate what they read; and there is no guarantee that they will always be successful in trying to actively and systematically understand and evaluate what they read. Critical thinking is an approach to problems, not a specific procedure that always produces a correct answer.

Are You a Critical Thinker?

When you are faced with studying your psychology textbook, do you behave like a critical thinker? Please rate each of the following statements on a scale from −3 (strongly disagree) to +3 (strongly agree):

_____ I should carefully read every word.
_____ I should expect that hard work will improve my performance.
_____ I should focus on the terms and facts that the author presents.
_____ I should be able to recite the author's major conclusions.
_____ I should assume the author is an expert.
_____ I should use a foolproof system for studying.

If you are like most introductory psychology students, you agree with these statements. However, we would like to suggest that each of these statements reflects a potential misconception that could block your efforts to become a critical thinker.

First, many students believe that their primary job is to carefully read every word. However, as we indicated in the first part of our definition of critical thinking, carefully reading each word is just the beginning. A critical thinker must be actively involved in making sense out of the material. You need to see yourself as a "sense maker," not just a "word processor."

Second, many students believe that there is some magic guarantee of success. A common refrain following poor performance on an exam is ". . . but I worked very hard." As suggested by the second part of the definition, hard work without a systematic approach may be wasted work. You need to become a "smart worker," not just a "hard worker."

Third, many students believe that the main point of psychology is to learn to give the definitions of certain key terms and to recite facts about certain key topics. However, as we indicated in the third part of our definition, critical thinkers focus on an entirely different aspect of what the author is saying, namely the author's arguments. You need to see yourself as an "argument detector," not just a "fact collector."

3

Fourth, many students focus on the *product*—that is, the final conclusion—of the author's argument rather than on the *process* the author used for arriving at the conclusion. However, the fourth part of our definition makes clear that a critical attitude is at the heart of critical thinking. When it comes to psychological arguments, you need to focus on process rather than product.

Fifth, many students assume that the author is an infallible expert whose wisdom is to be memorized. However, according to the fifth part of our definition of critical thinking, you need to be a critical consumer of psychological information. When it comes to the science of psychology, it is better to question authority than to accept everything you read.

Sixth, many students harbor the belief that there is some foolproof system (usually involving almost no effort) that will make learning easy. However, the sixth part of our definition suggests that there is no shortcut to becoming a critical thinker. You need to see critical thinking as a goal that needs to be adjusted to specific situations rather than as a well-specified procedure that always works the same way.

What's missing in your current approach to studying psychology? If you are like most introductory psychology students, you need to become a more active and critical reader who can identify, analyze, and evaluate arguments. This book is intended to help you achieve this goal.

How Can You Learn to Become a Critical Thinker?

Learning to become a critical thinker involves three levels of change: an affective change, a cognitive change, and a behavioral change.

Affective change. The first change involves your feelings about the need for critical thinking. Your initial asset is your awareness of the need to be a critical thinker and your motivation to try out the methods suggested in this book. This will take commitment to becoming a critical thinker as well as the realization that you may need to modify our methods to fit your specific needs. Chapter 1 of this book focuses on changing your feelings about the importance of critical thinking.

Cognitive change. The second change involves acquiring specific knowledge about strategies for critical thinking. Chapters 2 and 3 of this book present some specific strategies for how to improve your critical thinking in psychology. Our main focus is on strategies for understanding and evaluating arguments in your psychology textbook.

Behavioral change. The third change involves your behavior when you read a psychology textbook. The exercises in this book, particularly chapters 4 through 8, are intended to help you apply your knowledge of critical thinking strategies to the outlining of your textbook.

In short, becoming a critical thinker involves your heart, your head, and your hand. You need to develop a commitment to critical thinking (affective change), to acquire specific knowledge about how to be a critical thinker (cognitive change), and to actually apply that knowledge when you read your psychology textbook (behavioral change).

Summary

A successful psychology student needs to learn both specific information about psychology and techniques for how to think critically in psychology.

Critical thinking in psychology is an active and systematic attempt to understand and evaluate arguments.

Becoming a critical thinker involves affective, cognitive, and behavioral changes.

Chapter 2

Arguments: The Basic Units of Critical Thinking in Psychology

What Is an Argument in Psychology?

When you look at a page of your psychology textbook, what do you see? If you are like most students in introductory psychology courses, you see lots of words and perhaps some illustrations. Sometimes all the words and pictures on the page get in your way, like not being able to see the forest for the trees. The purpose of this book is to help you look deeper when you read your textbook, so that you are able to find out what the author is talking about. In this chapter you will learn how to identify *arguments*.

The word "argument" may make you think of two people shouting at one another, perhaps even throwing pots and pans against the walls as they continue shouting. Alternatively, you might think of someone trying to persuade you to do something. However, we use the word in a different way in this book. An *argument in psychology* consists of an assertion along with empirical evidence and a theoretical explanation for the assertion. In other words, an argument is a statement describing the world (assertion) along with empirical support (evidence) and theoretical support (explanation) for the statement.

Psychology is full of arguments, including arguments about the role of nature and nurture in human development, the characteristics of the human information processing system, and where prejudice comes from. In becoming a critical thinker, your first step is to identify the arguments that the author presents.

What Are the Parts of an Argument?

As you can see in figure 2.1, a psychological argument generally consists of three parts:

(1) an *assertion*—that is, a general description of the characteristics of one or more things or the relation between two or more things,

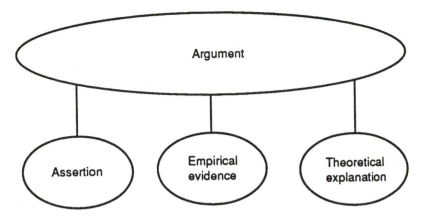

Figure 2.1. Three Parts of an Argument

(2) *empirical evidence*—that is, specific observations that support or refute the assertion, and
(3) *theoretical explanation*—that is, a hypothesized mechanism or model that logically justifies or refutes the assertion.

For example, suppose your friend claims that your university discriminates against unattractive people in its hiring practices. She provides survey results showing that 25% of the people in your state are unattractive but only 2% of the employees at the university are unattractive. She concludes that personnel interviewers tend to trust and believe attractive job candidates more than unattractive ones.

Let us ask you a few questions about your friend's argument. Please take the time to write down an answer to each question, using your own words.

First, what is the assertion?

_____

_____

_____

Second, what is the evidence?

_____

_____

_____

Third, what is the explanation?

_____
_____
_____

    In this example, the assertion describes a relationship between two things—degree of attractiveness and probability of being employed by the university. In particular, the assertion is that the less attractive a person is, the less likely he or she is to be hired. Notice that we have slightly changed your friend's wording that the university "discriminates against unattractive people in its hiring practices" so that we can focus on relationships between things. In this example, the supporting empirical evidence is the survey result that 2% of the employees are unattractive but 25% of the people in the state are unattractive. Finally, the theoretical explanation concerns a mechanism, that is, that personnel interviewers believe attractive candidates more than they believe unattractive ones.

    Can you criticize your friend's argument? Please list as many things as you can think of that are wrong with the assertion, evidence, and explanation.

What's wrong with the assertion?

_____
_____
_____

What's wrong with the evidence?

_____
_____
_____

What's wrong with the explanation?

_____
_____
_____

    When students answer these questions, they generate a number of high quality critical answers. For example, a major problem that students often raise about the assertion concerns how to define and measure attractiveness. Similarly, critical students typically have questions about how the evidence was obtained. For example, was the survey based on samples of workers in the state and samples of university workers who were really representative of those groups? Finally, in criticizing the explanation, students typically point out that there are alternative theories that could account for the evidence—such as the possibility that when people become competent workers they also tend to make themselves look more attractive by using good grooming and fashion techniques. Another alternative explanation may be that the pool of applicants to the university contains a high percentage of attractive people and a very low percentage of unattractive people. How do your answers compare with these?

Types of Assertions in Psychology

Let's take a closer look at assertions. *An assertion is a general description of something or the relation among things.* A careful inspection of your textbook reveals that psychological assertions generally come in one of four types or formats. As you read your text, you need to be on the lookout for assertions. You will often need to recast the author's wording of the assertions into one of the four kinds summarized in table 2.1 and described as follows.

Assertion Type 1: A and B are related. The first type of assertion describes a relation between two or more things. In a relational assertion, *A and B are related.* For example, someone might assert that the best predictor of a student's grades in college is his or her grades in high school. In this example, the things that are alleged to be related are "grades in college" and "grades in high school."

Assertion Type 2: A causes B. The second type of assertion describes a special kind of relation between two or more things, namely, that one thing helps to cause the other thing. In short, *A is the cause, and B is the effect.* For example, a causal assertion is that parents who were abused as children are more likely to abuse their own children. In this case, the cause is "being abused as a child," and the effect is "abusing children as a parent."

Assertion Type 3: P is a property of A. The third type of assertion describes one or more characteristics or properties of a thing. In property assertions, *P is a characteristic, and A is a thing.* For example, someone might assert that humans can actively think about only one thing at a time. In this example, the property is "being able to think about one thing at a time," and the thing is "the human memory system."

Assertion Type 4: A differs from B on property P. The fourth type of assertion describes a specific relation between two or more things, namely, that one thing is different from another thing in some way.

9

Table 2.1. Four Types of Assertions

Type 1:   A and B are related.
Type 2:   A causes B.
Type 3:   P is a property of A.
Type 4:   A differs from B on property P.

In a differential assertion, *A and B differ with respect to some characteristic or property.* For example, someone might assert those tired old clichés that boys are better in math than girls, and girls are better in language arts than boys. In this example, there are two differential assertions: (1) that boys and girls differ with respect to mathematical ability, and (2) that boys and girls differ with respect to language ability.

Do these four types of assertions make sense to you? To test your understanding, we list several statements taken from psychology text-books. For each statement, write a "1" if it is a type 1 assertion, "2" if it is a type 2 assertion, "3" if it is a type 3 assertion, "4" if it is a type 4 assertion, and "NA" if it is not an assertion. Also, for each, rewrite the assertion in your own words according to one of the four formats you just read about. Where appropriate, underline and label *A, B,* and *P.* Go ahead and give it a try.

_____ 1. Apes can learn human language.
_____ 2. Hormones regulate sexual motivation.
_____ 3. Practicing meditation will lower your blood pressure.
_____ 4. Many soft drinks contain caffeine.
_____ 5. The probability of alcohol abuse corresponds to the amount of stress in a person's life.
_____ 6. The Graduate Record Exam (GRE) in Psychology predicts whether you will be successful at graduate work in psychology.
_____ 7. Individuals with schizoid personality disorder have little or no desire for friends.
_____ 8. Left-handed people are more creative than right-handed people.

10

9. In making a judgment about someone, people tend to weigh negative information more heavily than positive information.
10. Children of parents with college degrees tend to have better attitudes about school than children of parents who lack college degrees.

Here are our answers (indicated by italics):

Type 3   1. *Language ability* (P) is a property of *apes* (A).
Type 2   2. *Hormones* (A) cause *sexual motivation* (B).
Type 2   3. *Meditation* (A) causes *lower blood pressure* (B).
NA   4. This is a factual statement that does not involve human behavior.
Type 1   5. *Probability of alcohol abuse* (A) and *amount of stress* (B) are related.
Type 1   6. *Score on GRE* (A) is related to *success in graduate school* (B).
NA   7. This is a definition.
Type 4   8. *Left-handed people* (A) differ from *right-handed people* (B) on *degree of creativity* (P).
Type 3   9. *Weighing negative information more heavily than positive information* (P) is a property of *human judgment* (A).
Type 4   10. *Children of college-educated parents* (A) differ from *children of non-college-educated parents* (B) on *attitudes about school* (P).

How do your answers line up with ours? If you disagree with an answer, try to reconcile the disagreement by rereading the definitions of each type of assertion. Sometimes you may be able to state an assertion in several ways. For example, assertion 3 could be viewed as a causal assertion (meditation causes lowered blood pressure), as a relational assertion (amount of meditation and level of blood pressure are related), or as a property assertion (lower blood pressure is a property of people who meditate). The causal assertion (type 2) is the most appropriate form because the author's statement of the assertion emphasizes cause-and-effect. However, if another author emphasized the relation without emphasizing the cause-and-effect, then you would be correct in viewing the assertion as most closely matching type 1.

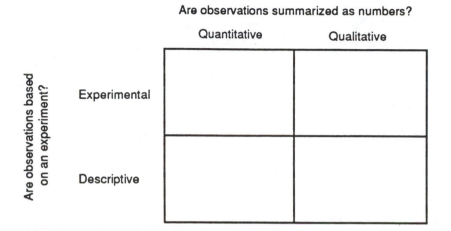

Figure 2.2. Four Types of Empirical Evidence

Types of Empirical Evidence in Psychology

The second component in a psychological argument is empirical evidence. The term *empirical* means observable. *Empirical evidence is observable data, such as the results of observing a natural setting or a laboratory experiment.* Examples include observing the behavior of children in a classroom or the results of an experiment in which students were taught under two different instructional methods. Furthermore, it is important to note that you can make observations in these situations without any special abilities. For example, the empirical data collected by one person should be the same as the data collected by another observer looking at the same situation.

In psychology, critical thinking involves considering how well the empirical evidence supports the assertion. To do this, you need to understand the kinds of empirical evidence that psychologists gather. Figure 2.2 summarizes four major types of empirical evidence used in psychology. When you come across empirical evidence in your textbook, you can ask yourself two questions. First, is the evidence measured in numbers (i.e., *quantitative*) or in words and pictures (i.e., *qualitative*)? Second, is the evidence based on observations in a natural setting (i.e., *descriptive*) or on the results of experiments (i.e., *experimental*).

As summarized in figure 2.2, four major ways to collect empirical evidence in psychology are the following.

Quantitative-descriptive evidence. This type of evidence is based on observations of the natural environment that are expressed in numbers, including survey results and regular test scores. For example, a researcher who is investigating social cooperation in learning might observe children in a classroom and count the number of times each student offers help to or asks for help from another student.

Qualitative-descriptive evidence. This type of evidence is based on observations of the natural environment that are not expressed in numbers, including clinical interviews and subject protocols. For example, a researcher might write a descriptive narrative of a child's social interactions with other students over the course of the school day.

Quantitative-experimental evidence. This type of evidence consists of the results of an experiment that are expressed as numbers. For example, a researcher might arrange one classroom so that children work in small groups and another classroom so that children work as a whole class. This is an experiment because the researcher is actively manipulating the environment, rather than just measuring it in some way. Further, suppose that the researcher counts the number of times each student offers or receives help in each of the two classrooms.

Qualitative-experimental evidence. This type of evidence consists of the results of an experiment that are not expressed as numbers. For example, a researcher may write narrative descriptions of a child who was assigned to a small group and a child who was assigned to a whole-class arrangement.

As you read your psychology textbook you should be on the lookout for each type of empirical evidence. For example, for each of the following pieces of evidence, write either "1" for quantitative-descriptive, "2" for qualitative-descriptive, "3" for quantitative-experimental, or "4" for qualitative-experimental.

_____ In a midwestern college, half the students were required to eat each of their meals in cafeterias that were painted "institutional green," and half were required to eat in cafeterias that were painted in soft pink, turquoise, and peach. Although the food served in the two cafeterias was identical, student ratings of the food in the institutional green cafeteria averaged 2.0, on a scale of 1 to 5 (with 1 as worst and 5 as best), whereas student ratings from the pastel cafeteria averaged 4.0.

13

____ An analysis of 100 drivers involved in serious traffic accidents revealed that 25% had recently suffered a financial or personal loss such as a divorce, death of a loved one, or loss of a job.

____ In an analysis of his problem-solving behavior for "2 + 4 = ___," a first grader reported, "I take one from the 4 and give it to the 2. Now, I have 3 plus 3, and I know the answer is 6."

____ One student was taught computer programming using the standard manual and another student was taught by allowing lots of hands-on experience. When solving a new programming problem, the traditionally taught student said, "I haven't had this yet," whereas the actively taught student said, "This is fun."

The correct answers for the above items are "3," "1," "2," and "4." Note that manipulations of the environment that produce evidence are *experiments,* whereas observations of naturally occurring events are *descriptions;* also note that when measurements involve numbers, they are *quantitative;* and when they do not, they are *qualitative.* When your textbook presents evidence, which of these four types is most often given?

Let's try some problems. For each of the following assertions, list examples of empirical evidence that you would like to have in order to test the assertion. You can be creative by making up evidence that would help to support or refute the assertion.

1. Grades in high school are the best predictor of grades in college.

   _____

   _____

   _____

2. If a person is abused as a child, the person will be abusive as a parent.

   _____

   _____

   _____

3. Humans are able to actively think about only one thing at a time.

   _____

   _____

   _____

4. Boys have more mathematical ability than girls, and girls have more language ability than boys.

_____

_____

_____

Here are some of the best answers that we generated.

1. For a large sample of students, get their high school grade point average and their college grade point average.
2. Survey a group of child-abusing adults and an otherwise equivalent group of non-abusing adults to see what percentage of each group was abused as children.
3. Give people several demanding cognitive tasks to do simultaneously (such as reciting the alphabet backward, writing numbers by 3s starting from 10, and tapping one's toe in conjunction with a series of tones that are presented) and see if they can do them.
4. Get scores on math and language ability tests for boys and girls who have had identical past experience with math and language.

How do your answers correspond to ours? As you can see, there are many different appropriate sources of empirical evidence.

Theoretical Explanations in Psychology

Now comes the last part of an argument. You have explored the first two parts, the assertion and the empirical evidence for the assertion. The third part of an argument is the theoretical explanation for the assertion. *The theoretical explanation contains a mechanism or principle that tells how something works; that is, it explains why the assertion happens.*

It is important for you to understand the distinction between description (i.e., the assertion) and explanation (i.e., the theoretical explanation). A description is a statement about the way the world is—such as A and B are related, A causes B, P is a property of A, or A and B differ on property P. An explanation is a statement about why the world is that way, usually consisting of a model with parts that interact. A model is a simplified description of some system: it contains parts that interact in lawful ways. For example, please take a moment to try to generate some possible theoretical explanations of four assertions.

15

1. Grades in high school are the best predictor of grades in college.
    _____
    _____
    _____

2. If a person is abused as a child, the person will be abusive as a parent.
    _____
    _____
    _____

3. Humans are able to think actively about only one thing at a time.
    _____
    _____
    _____

4. Boys have more mathematical ability than girls, and girls have more language ability than boys.
    _____
    _____
    _____

Here are some of the best answers that students have generated.

1. Intellectual ability theory: students with higher intellectual ability will achieve higher grades in both high school and college.
    Prior knowledge theory: students who learned more in high school will be better able to assimilate new knowledge in college.
2. Genetic theory: the tendency to be abusive is inherited.
    Role model theory: the tendency to be abusive is learned from parental role models.
3. Single channel for input: people can pay attention to only one input channel at a time.
    Single channel for output: people can control only one output channel at a time.
4. Genetic theory: mathematical and language ability are inherited and are sex-linked.
    Socialization theory: boys tend to have more experience with mathematical activities, whereas girls have more experience with language activities.

As you can see, each of these theories suggests some additional observations that you might want to make in order to test them further.

Each time you encounter an assertion and empirical evidence in your psychology textbook, it is important for you to be on the lookout for accompanying theories. The theoretical explanation may not always be presented, partly because psychology is still an emerging science.

Summary

Arguments are the basic unit of critical thinking in psychology. An argument consists of three parts: assertion, evidence, and explanation.

An assertion states a general relation between two things or the characteristics of one or more things. Four types of assertion are: A and B are related, A causes B, P is a property of A, A differs from B on property P.

Empirical evidence consists of specific observations that support or refute the assertion. Four kinds of empirical evidence are: quantitative-descriptive, qualitative-descriptive, quantitative-experimental, and qualitative-experimental.

Theoretical explanations are hypothesized mechanisms or principles that logically justify or refute the assertion.

Chapter 3

Strategies for Understanding and Evaluating Arguments

Introduction

Consider the following scenario: you are sitting at your desk with your psychology textbook open and you have allocated some time to study. It's just you and the psychology book. What should you do?

In this chapter you will learn two basic strategies that you can use when you read your psychology textbook: strategies for *understanding* what you read and strategies for *evaluating* what you read.

How to Understand an Argument

In order to delve more deeply into what you read, you must be sure that you understand the arguments that the author presents to you. Table 3.1, labeled "Do I Understand the Argument?," presents some questions that can help you understand arguments in your psychology textbook. Let's look at each of the three steps for understanding.

Understanding the assertion. The first step in understanding an argument is to understand what the argument is about; that is, you must be able to identify the assertion that the author is making. This is not always an easy task, because the author may not always present the assertion clearly. For example, you may need to translate the author's wording of an assertion into one of the four formats you saw in chapter 2—A is related to B, A causes B, P is a feature of A, and A differs from B on property P. To help you understand the assertion, the survey asks you to state the assertion, restate the assertion in your own words, and give a concrete example.

Table 3.1. Do I Understand the Argument?

1a.   *Can I state the assertion (and underline the key terms)?*

Copy or paraphrase the essential assertion from the text, such as a statement saying that there is a certain relationship between two things (or factors) or that a thing (or factor) has a certain characteristic. Underline the key terms; that is, underline the names of the things (or factors).

1b.   *Can I put the assertion in my own words?*

Translate the assertion into your own words without using technical jargon; try to word your statement in a way that the average "person-on-the-street" could understand.

1c.   *Can I give a concrete example of the assertion (including underlining and labeling examples of the key terms)?*

Provide a concrete example—either from the passage or made-up by you—that corresponds to the assertion. Underline and label the parts of your example that correspond to each of the key terms.

2.   *Can I state the empirical evidence?*

For each piece of empirical evidence, provide the following information.
Name: Give the name(s) of the researchers and the date of publication, if available.
Method (including underlining and labeling the independent and dependent variables): Give a brief summary of the method, including a clear statement of the independent variable and the dependent variable.
Results: Give a brief summary of the results (statistics are not needed).
Conclusion: State how well the assertion is supported, or the degree to which the assertion is supported.

3.   *Can I state the theoretical explanation?*

State the theoretical mechanism that accounts for the results, if one is stated in the passage. Explain why the assertion is correct, based on a theory rather than based on data.

Understanding the empirical evidence. The second step is to identify empirical evidence. The most common empirical evidence is a research study. In general, a research study has three parts: (1) method, (2) results, and (3) conclusion. You should be able to summarize each of these parts in one or two sentences.

In *summarizing the method,* you should describe the participants, materials, and procedures used in the study. In your description you should be sure to clearly specify the *independent variable(s)*—i.e., what was manipulated—and the *dependent variable(s)*—i.e., the resulting behavior that was measured. For example, consider the following summary: "College students read a 500-word essay that either contained illustrations (experimental group) or did not (control group), and took a recall test." In one sentence it gives the independent variable (presence or absence of illustrations) and the dependent variable (performance on a recall test) as well as a brief summary of the participants (college students), the materials (500-word essays), and the procedure (subjects read and were tested).

In *summarizing the results,* you should describe the main finding(s) in words by telling what was measured and for which treatment groups, such as "The experimental group recalled approximately twice as many idea units from the essay as the control group." You do not need to report specific numbers or specific statistical tests, but any result you report should be statistically significant.

In *summarizing the conclusion,* you should describe how the results support or refute one or more theories. For example, consider the conclusion: "The results are consistent with the idea that readers used the illustrations as cues for how to organize the material into meaningful units." A conclusion should do more than repeat the results; it should tell how the results relate to the theory.

Understanding the theoretical explanation. The third step involves explaining the underlying mechanism, that is, spelling out a logically coherent theory. For example, what are the intervening processes and mechanisms that account for the description of the world given in the assertion?

To make sure that you can carry out these three steps, please read the passage on *Positive Reinforcement* in box 3.1 and then fill out your answers to the questions in table 3.2. This process should take at least 15 to 20 minutes. Although strategies for understanding require far more time than it takes to "read every word," the effort will help you to achieve a higher level of conceptual mastery of the material that will help you answer questions about the material.

Box 3.1. Positive Reinforcement

If you started parking your car in a new parking space marked "For the President Only" and your car was towed away every day as a consequence, you would probably stop parking there after a while. Similarly, if you moved to a new seat in class and suddenly an interesting, attractive person started talking with you, you would probably choose to sit in that seat for the remaining classes of the semester. To a great extent, people increase or decrease the frequency with which they do things depending on the consequences of their actions.

In *positive reinforcement,* the consequences of a behavior are positive, so the behavior is engaged in more frequently. Simply stated, we say that positive reinforcement has occurred whenever consequence of behavior leads to an increase in its probability of occurrence.

In the early 1960s, a team of preschool teachers helped a young girl overcome her shyness in what has become a widely cited example of the principle of positive reinforcement (Allen, Hart, Buell, Harris, & Wolf, 1964). The teachers were worried because the girl spent little time playing with the other children and too much time with her adult teachers. They decided to encourage peer play through positive reinforcement. They knew that she enjoyed receiving praise from the teachers, so they decided to praise her only when she was playing with another child. This made the consequence of playing with other children a positive one. The results of this use of positive reinforcement are shown in figure 3.A. To be able to evaluate the results of their positive reinforcement program, the teachers first counted the frequency with which the little girl interacted with other children and with adults before anything was done to help her. Then they started to positively reinforce (praise) her for playing with other children, but otherwise they paid very little attention to her (so that she would get positive reinforcement from her teachers only for playing with peers). As can be seen in the second segment of figure 3.A, the little girl's frequency of playing with peers increased markedly when the teachers reinforced her. To be sure that the positive reinforcement was responsible for the changes and not some other factor, the teachers stopped reinforcing her for playing with peers in the third segment (the *reversal* phase) of figure 3.A, and then reinforced her again during the fourth phase. As can be seen, the frequency of peer play dropped when the positive reinforcement was discontinued, but increased again when it was resumed in the fourth phase. Thus, the teachers were able to intentionally teach a more adaptive pattern of playing to this child by using positive reinforcement.

Many other applications have been made of this principle, ranging from teaching hospitalized schizophrenic adults more normal patterns of behavior to teaching employees to reduce the amount of damage sustained when sorting boxes for airfreight delivery (Rimm & Masters, 1979). In each case, the behavior that becomes more frequent is termed the *operant response* and the positive consequence of that response is called the *positive reinforcer.*

Figure 3.A. *Increasing the amount of time that a child spends playing with other children through the use of positive reinforcement.*

I do not want to give the impression through this discussion that positive reinforcement is something that only occurs when it's intentionally arranged. The natural consequences of our behavior can be reinforcing as well. For example, we learn that some ways of interacting with our friends or supervisors naturally lead to happier relationships, and that is positively reinforcing. We are always affected by the consequences of our behavior and hence are always in the process of learning to adjust to our world through positive reinforcement.

Table 3.2. Do I Understand the Argument?

1a.   *Can I state the assertion (and underline the key terms)?*

1b.   *Can I put the assertion in my own words?*

1c.   *Can I give a concrete example of the assertion (including underlining and labeling examples of the key terms)?*

2.   *Can I state the empirical evidence?*
Name:

Method (including underlining and labeling the independent and dependent variables):

Results:

Conclusion:

3.   *Can I state the theoretical explanation?*

Table 3.3. Karin's Understanding of Positive Reinforcement

1a.   *Can I state the assertion (and underline the key terms)?*

       Positive consequences (i.e., positive reinforcement) of a *behavior* (i.e., operant response) cause an increase in its probability of occurrence.

1b.   *Can I put the assertion in my own words?*

       If I get a reward for doing something—like praise or a prize—then I'm more likely to do it again.

1c.   *Can I give a concrete example of the assertion (including labeling examples of key terms)?*

       If I get a *good grade* (positive reinforcer) after *studying in front of the television for a test* (operant response) then I'm more likely to study that way the next time.

2.   *Can I state the empirical evidence?*

Name: Allen et al (1964)

Method: Teacher observes that a preschool child spends much time with adults and less time with other children (dependent variable = amount of time spent with other children) so she praises the child for playing with other children but not for interacting with adults (independent variable = presence of praise for interacting with children).

Results: Child increases amount of time spent with other children. The change occurred as soon as the praise started.

Conclusion: Positive reinforcement (praise) causes an increase in the reinforced behavior (interacting with children).

3.   *Can I state the theoretical explanation?*

       Each time a learner receives positive reinforcement following a behavior in a certain situation, the learner automatically strengthens the association between that behavior and the situation. (Note: The underlying theoretical mechanism of "strengthening associations" is discussed in a previous section of the chapter.)

Becoming a critical thinker requires that you take the time to understand the author's argument. Let's see how your answers to the survey compared to those given by a highly successful psychology student whom we will call Karin. Table 3.3 gives Karin's answers to the survey. Please compare Karin's answers in table 3.3 to your answers in table 3.2; summarize the major differences in table 3.4. This exercise should take about 5 to 10 minutes and should help you to adjust your performance.

Table 3.4. Differences between What Karin and I Wrote

1a.   Can I state the assertion?

1b.   Can I put the assertion in my own words?

1c.   Can I give a concrete example of the assertion?

2.   Can I state the empirical evidence?
        Name:

        Method:

        Results:

        Conclusion:

3.   Can I state the theoretical explanation?

Table 3.5. Can I Evaluate the Argument?

---

1. *Can I evaluate the assertion?*

    Critically comment on each of the following aspects of the assertion.

    Clarity: Tell whether or not each term in the assertion is clearly defined and easy to measure.

    Testability: Tell how well the assertion could be tested and, in particular, whether or not it would be possible to prove it wrong.

    Importance: Tell how important or significant the assertion is, i.e., whether or not the issue is worth studying.

2. *Can I evaluate the empirical evidence?*

    For each piece of empirical evidence, provide critiques of the method, results, and conclusion.

    Name: Give the name of the researcher(s) and date of publication, if available.

    Method: Identify possible sources of bias in the method, including lack of experimental control.

    Results: Can you trust the results? Is there another way to look at the data?

    Conclusion: Does the conclusion follow logically from the results?

3. *Can I evaluate the theoretical explanation?*

    Is there an alternative explanation for the evidence? If so, describe the alternative explanation and describe how it can account for the results.

How to Evaluate an Argument

Now, let's get really critical. Table 3.5, "Can I Evaluate the Argument?," includes three major sets of questions.

Evaluating the assertion. The first step is to evaluate the assertion by judging its clarity, testability, and importance. To evaluate clarity, find each key term in the assertion and determine whether you would have any difficulty in defining and measuring it. If you cannot define the terms or figure out how to measure them, the assertion is too vague. To evaluate testability, ask yourself whether it would be possible to find empirical evidence that could disprove the assertion. If there is no way to prove that the assertion is false, it is not a testable assertion. To evaluate the importance of the assertion, you

26

must ask yourself whether the issue being tested makes a contribution to our understanding of human behavior or mental life. Not all assertions are equally worthy of study.

Evaluating the empirical evidence. The second step is to evaluate the empirical evidence by looking for flaws in the method, results, or conclusion. To evaluate the method of study, look for mistakes in the way the study was conducted. One common source of bias is *overlooking the real independent variable,* that is, failing to control for other potential independent variables. For example, a kindergarten teacher who uses a special art program finds that children show an improvement in their artwork from the beginning of the year to the end of the year. She concludes that she has taught the children to be better artists, but, in fact, the real independent variable may be time; children may have matured over the year so they have better control over their fine motor movements. The teacher needs to compare the gains of her students to those in a traditional kindergarten art program. Another common error is *inadequate sampling,* that is, failing to select participants or materials adequately. For example, in conducting an election survey, inadequate sampling would occur if you randomly selected names from the telephone book. This sample would not include low-income voters who do not have telephones and hence would not provide a representative survey. Try to detect as many methodological flaws as you can.

To evaluate the results, ask yourself whether data were ignored that conflicted with the predictions, or whether data were scored in a way that favored the predictions. For example, in observing students' classroom behavior, a teacher might be more aware of productive performance from "good" students than from "poor" students. If you find that the data are not selected, scored, or analyzed fairly, you have detected a flaw in the data.

To evaluate the conclusion, ask whether the explanation that is given actually follows from the data. Ask yourself whether there is a strong logical connection between the conclusion and the data. If not, you have found another flaw.

Evaluating the theoretical explanation. The third step is to evaluate the theoretical explanation by determining whether another explanation would be better. Can you think of another explanation for the data that can account for the data?

To make sure you can follow these steps, please answer the questions in table 3.6 concerning the positive reinforcement passage.

27

Table 3.6. Can I Evaluate the Argument?

---

1.   *Can I evaluate the assertion?*
       Clarity:

       Testability:

       Importance:

2.   *Can I evaluate the empirical evidence?*
       Name:

       Method:

       Results:

       Conclusion:

3.   *Can I evaluate the theoretical explanation?*

This exercise should take about 15 to 20 minutes. There is no one right answer so try to generate as many critical comments as you can think of.

Once you have filled out table 3.6, please look at table 3.7, which gives Karin's answers. In table 3.8, carefully list the differences between your answers and Karin's answers. This exercise should take 5 to 10 minutes and should help you to adjust your strategies for evaluation.

Table 3.7. Karin's Evaluation of Positive Reinforcement

1. *Can I evaluate the assertion?*

    Clarity: The definition of operant response is clear but the definition of positive reinforcer is circular.

    Testability: If a positive reinforcer does not work, it is possible to claim that it was not appropriate for the learner, e.g., praise does not appeal to some children.

    Importance: It is very important to find reliable methods of influencing and possibly controlling the frequency of certain types of behavior.

2. *Can I evaluate the empirical evidence?*

    Name: Allen et al (1964)

    Method: There was only one subject. The adults involved had a strong social influence on the child. How aware was the child of the adult expectations?

    Results: Why was there an immediate increase in behavior rather than a gradual change? This is inconsistent with the idea that learning involves gradual strengthening of associations.

    Conclusion: The results do indicate a relationship between positive reinforcement and behavior change, but the reason for the relationship is not clear.

3. *Can I evaluate the theoretical explanation?*

    Since there is not a strong theoretical statement given in this part of the text, I had to find it in an earlier section of the chapter. An alternative explanation is that the child interprets the teacher's praise as an instruction from the teacher to spend more time with other children; in other words, the child's cognitive interpretation of the reinforcement is what makes it have an effect.

Table 3.8. Differences between What Karin and I Wrote

1.   *Can I evaluate the assertion?*
       Clarity:

       Testability:

       Importance:

2.   *Can I evaluate the empirical evidence?*
       Name:

       Method:

       Results:

       Conclusion:

3.   *Can I evaluate the theoretical explanation?*

Summary

Understanding an argument involves being able to summarize the assertion using one of the four formats in chapter 2; the empirical evidence, including the method, results, and conclusion from each research study; and the theoretical explanation, including any underlying principles or mechanisms.

Evaluating an argument involves judging the clarity, testability, and importance of the assertion; identifying flaws in the method, results, or conclusion of research studies; and generating alternative explanations.

Part 2

Applying Strategies for Critical Thinking

Chapter **4**   **Critical Thinking Exercise A**

Social Psychology

Do You Understand the Argument?

Now, you are ready to apply the strategies you have learned to real passages from introductory psychology textbooks. Please carefully read the passage on *Cognitive Dissonance* that is presented in box 4.1. Then, turn to table 4.1 and fill in the spaces, indicating how you summarize the assertion, evidence, and explanation given in the passage. If you have difficulty, you should reexamine table 3.1.

▓▓▓   **Box 4.1.** Cognitive Dissonance

Cognitive Dissonance Theory of Attitude Change

Leon Festinger (1957) proposed the theory of **cognitive dissonance** to explain the tendency of attitudes to sometimes shift to be consistent with behavior. This theory, which has sparked some of the greatest controversy and most interesting research in social psychology, states that inconsistencies between attitudes and behavior are uncomfortable. This discomfort motivates people to do what they can to reduce the discomfort, or *dissonance.*

For example, if you smoke cigarettes (behavior), and you know that cigarette smoking is the leading cause of lung cancer and other horrible diseases (attitude), your behavior and attitude would be inconsistent, which would produce dissonance. Dissonance theory predicts that either your attitude or your behavior will change to reduce the dissonance (see figure 4.a). You could change the behavior and quit smoking, but that is often very difficult for smokers. Unfortunately, human beings usually reduce dissonance in the easiest way possible. In this case, it might be easier to change your attitude toward smoking. How many smokers do you know who say that the research linking

32

cancer and smoking is faulty, or who say, "So what if smoking causes cancer, I'll die of smoething anyway?" These are irrational and self-defeating arguments, but they are sadly effective ways of reducing dissonance.

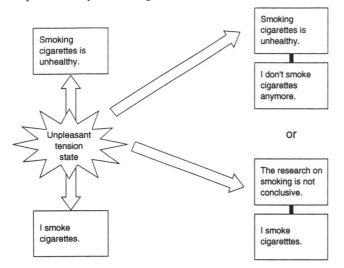

Figure 4.A. *Cognitive Dissonance. If attitudes and behavior are inconsistent, dissonance is aroused. To reduce this unpleasant state, either attitudes or behaviors change to be consistent with one another.*

Festinger and other social psychologists tested the theory of cognitive dissonance in a number of experiments. One of the best known (Festinger & Carlsmith, 1959) involved asking subjects to perform a boring spool-stacking task for an hour. Afterward, the subjects were each asked to tell the next subject that the task was an interesting one. Half of the subjects were offered $20 to say that the task was interesting, and half were offered $1. A third group of subjects stacked spools, but were not asked to say anything to the next subject. Later, all of the subjects were asked how interesting they really thought the task was.

Which group do you think reported the most favorable attitude toward the task? Perhaps surprisingly—but, just as predicted by cognitive dissonance theory—the most positive attitudes were expressed by the group offered only $1. The group offered $20 was not placed in a state of dissonance: "The task was really boring, but I'll lie to the next person to get the $20." The group offered $1 was placed in a state of dissonance, however; there was no good explanation for their stating an opinion about the task that was inconsistent with their attitude so their attitudes proved to be more consistent with their behavior.

A more recent study shows that cognitive dissonance is relevant to more than our attitudes about stacking spools (Cooper & Mackie, 1983). University students who supported Ronald Reagan in the 1980 presidential election were the subjects in this experiment. They were asked to write an essay favoring an issue that Ronald Reagan opposed (federally sponsored health care) or an essay supporting the candidacy of Jimmy Carter. Half of the subjects were given very little choice about writing the essays. This group of subjects could be expected to experience little cognitive dissonance ("I wrote an essay that was contrary to what I believe, but I had to do it"). The other group of subjects were given much more choice as to whether or not they would write the essay. These subjects were likely to experience considerable cognitive dissonance, since they wrote an essay favoring something that they opposed even though they did not have to do so. As Festinger would have predicted, attitudes toward Jimmy Carter and federally sponsored health care changed very little in the low-cognitive dissonance group (the subjects given little choice about writing the essays), but changed significantly more in the high-cognitive dissonance group (the ones given more choice). When the behavior of writing the essays created cognitive dissonance, attitudes changed to be more consistent with the behavior.

It's a little scary, isn't it? You probably thought your attitudes were always thoughtfully arrived at and based on reality. If Festinger is right, they may sometimes reflect nothing more than an escape from cognitive dissonance.

From Benjamin B. Lahey, *Psychology,* 5th ed. Copyright © 1995 Wm. C. Brown Communications, Inc., Dubuque, Iowa. All Rights Reserved. Reprinted by permission.

Table 4.1. Do I Understand the Argument?

1a.   *Can I state the assertion (and underline the key terms)?*

1b.   *Can I put the assertion in my own words?*

1c.   *Can I give a concrete example of the assertion (including underlining and labeling examples of the key terms)?*

2.   *Can I state the empirical evidence?*
        Name:

        Method (including underlining and labeling the independent and dependent
               variables):

        Results:

        Conclusion:

3.   *Can I state the theoretical explanation?*

Table 4.2. Can I Evaluate the Argument?

---

1.   *Can I evaluate the assertion?*
      Clarity:

      Testability:

      Importance:

2.   *Can I evaluate the empirical evidence?*
      Name:

      Method:

      Results:

      Conclusion:

3.   *Can I evaluate the theoretical explanation?*

Can You Evaluate the Argument?

Next, fill in the spaces in table 4.2, indicating how you criticize the assertion, evidence, and explanation given in the text. If you have difficulty, you should reexamine table 3.5.

How Well Did You Do?

Were you able to understand and evaluate the argument presented in the *Cognitive Dissonance* passage? Take a moment to compare your summaries with those given in table 4.3 and table 4.4. For each number 1 through 3 in table 4.3 and in table 4.4, write "yes" if your answer was better or equivalent to our answer and write "no" if your answer was worse than our answer. Summarize your scoring by writing "yes" or "no" for each of the following questions:

___ Did I understand the assertion?
___ Did I understand the evidence?
___ Did I understand the explanation?
___ Could I evaluate the assertion?
___ Could I evaluate the evidence?
___ Could I evaluate the explanation?

If you answered "no" for any question, use the following space to write down a revised answer.

Table 4.3. Understanding Cognitive Dissonance

1a. *Can I state the assertion (and underline the key terms)?*

We need consistency between what we believe and what we do. When individuals feel *inconsistency* between their *attitudes* and *behavior,* they change their attitudes in order to reduce the discomfort they are sensing.

1b. *Can I put the assertion in my own words?*

When I am uncomfortable about a mismatch between what I am doing and what I believe, then I will reduce this discomfort by changing my beliefs to be more consistent with what I am doing.

1c. *Can I give a concrete example of the assertion (including labeling examples of key terms)?*

If *I accept an invitation* (behavior) to play tennis at a country club whose membership *policy I believe to be discriminatory* (attitude), then I will feel discomfort from the *mismatch* (inconsistency) between what I believe and what I am doing. To relieve this feeling, I will change my attitude about the club's policies.

2. *Can I state the empirical evidence?*

Name: Festinger & Carlsmith (1957)

Method: Three groups performed a boring task (stacking spools) for an hour and reported to the next subject on how they found the activity. One group was offered $20 to say that the job was interesting, another group was offered $1 to say it was interesting, and the third group was not asked to say anything. Later the subjects rated how much they really liked or did not like the task. Amount of payment for saying the task was interesting is the *independent variable*; the rating the participants gave the task is the *dependent variable.*

Results: Subjects who received $1 to say the task was interesting rated the task as more interesting than subjects who were paid $20.

Conclusion: Subjects who were paid $1 needed to justify why they said the task was interesting to the next subject, so they changed their attitudes about the interestingness of the task. Subjects paid $20 could justify their behavior on the grounds that they had been paid a fair amount of money, so they did not have to change their attitudes.

3. *Can I state the theoretical explanation?*

When there is inconsistency between what we say and what we do, we feel discomfort. To reduce the discomfort, we change our attitudes to conform more closely to our actions.

Table 4.4. Evaluating Cognitive Dissonance

---

1. *Can I evaluate the assertion?*
    Clarity: There is not much detail about how to measure the level of dissonance (discomfort). The prediction of a change in attitude is clear.
    Testability: The researchers assume that spool stacking is equally boring for all subjects.
    Importance: How attitudes change is an important issue.

2. *Can I evaluate the empirical evidence?*
    Name: Festinger & Carlsmith (1957)
    Method: The situation was contrived and may not be representative of attitude change in the real world. Subjects may have understood the point of the experiment. How did the experimenter get the subjects to do something they did not want to do—namely, tell a lie? Some subjects may have actually found the task to be interesting.
    Results: Ratings are a dubious measure. A person's rating may not really reflect his or her attitude.
    Conclusion: We do not know whether more subjects in the $1 group than the $20 group entered the experiment with the attitude that stacking spools was interesting. How do we know that the subjects in the $1 group really thought the task was interesting?

3. *Can I evaluate the theoretical explanation?*
    The key theoretical mechanism is dissonance (discomfort) reduction, but there is really no way to measure it. The ratings of the $1 group may be due to wanting to be positive or not wanting to appear stupid in front of the experimenter.

Developmental Psychology

Do You Understand the Argument?

Now, you are ready to apply the strategies you have learned to real passages from introductory psychology textbooks. Please carefully read the passage on *Maturation* that is presented in box 5.1. Then, turn to table 5.1 and fill in the spaces, indicating how you summarize the assertion, evidence, and explanation given in the passage. If you have difficulty, you should reexamine table 3.1.

Can You Evaluate the Argument?

Next, turn to table 5.2 and fill in the spaces, indicating how you criticize the assertion, evidence, and explanation given in the text. If you have difficulty, you should reexamine table 3.5.

▓▓▓   **Box 5.1.** Maturation

In the study of development, the most important aspect of nature (biological factors) is *maturation*. This term refers to systematic physical growth of the nervous system and other bodily structures. A primary question for the psychologist specializing in the study of development is, "How much of the change that we see occurring with age is the result of physical maturation rather than increased experience?"

      Although *both* experience and maturation are important roles in most developmental change, maturation is surprisingly important in many specific contexts. For example, experience obviously plays an important role in toilet training—children must be taught to use the toilet—but maturation also plays a key role. According to research, successful toilet training is difficult for most children before the age of 24 months. They are simply not maturationally ready to learn that task. After 24 months of age, however, most children learn rapidly.

      An experiment using identical twin boys illustrates the idea of maturational readiness beautifully. Toilet training was begun for one boy, Hugh, when he was only 50 days old. Training for Hilton, the other twin, did not begin until 700 days of age (almost 2 years). As shown in figure 5.A, Hugh made no real progress until about 18 months of age, while Hilton's progress was rapid from the beginning. Both children learned, but only when they reached the proper level of maturation (McGraw, 1940).

Figure 5.A. *The importance of maturational readiness is shown in McGraw's (1940) study of toilet training twin boys named Hugh and Hilton. Although Hugh's training was begun at 50 days of age, no progress was achieved until he was about 650 days of age. In contrast, Hilton's progress was rapid because training was begun when he was maturationally ready (McGraw, 1940).*

Table 5.1. Do I Understand the Argument?

1a.   Can I state the assertion (and underline the key terms)?

1b.   Can I put the assertion in my own words?

1c.   Can I give a concrete example of the assertion (including underlining and labeling examples of the key terms)?

2.   Can I state the empirical evidence?
Name:

Method (including underlining and labeling the independent and dependent
        variables):

Results:

Conclusion:

3.   Can I state the theoretical explanation?

Table 5.2. Can I Evaluate the Argument?

1. *Can I evaluate the assertion?*
   Clarity:

   Testability:

   Importance:

2. *Can I evaluate the empirical evidence?*
   Name:

   Method:

   Results:

   Conclusion:

3. *Can I evaluate the theoretical explanation?*

How Well Did You Do?

Were you able to understand and evaluate the argument presented in the *Maturation* passage? Take a moment to compare your summaries with those given in table 5.3 and table 5.4. For each number 1 through 3 in table 5.3 and table 5.4, write "yes" if your answer was better or equivalent to our answer and write "no" if your answer was worse than our answer. Summarize your scoring by writing "yes" or "no" for each of the following questions:

    ___   Did I understand the assertion?
    ___   Did I understand the evidence?
    ___   Did I understand the explanation?
    ___   Could I evaluate the assertion?
    ___   Could I evaluate the evidence?
    ___   Could I evaluate the explanation?

If you answered "no" for any question, use the following space to write down an improved answer.

Table 5.3. Understanding Maturation

1a.   *Can I state the assertion (and underline the key terms)?*

   Genetically determined growth of the nervous system and other bodily structures is related to the *development of behavior within a child.*

1b.   *Can I put the assertion in my own words?*

   Certain aspects of a child's behavior and learning can occur only when the brain and body are ready.

1c.   *Can I give a concrete example of the assertion (including labeling examples of the key terms)?*

   If I instruct a child on how to *ride a two-wheel bike* (development of behavior), she might not be able to ride it because her *sense of balance* (genetically determined growth) has not developed enough.

2.   *Can I state the empirical evidence?*

   Name: McGraw (1940)

   Method: Genetically identical twins, Hugh and Hilton, were toilet trained in different ways. Hugh's training started when he was 50 days old, and Hilton's started when he was 700 days old. The *independent variable* is the age at which toilet training began, and the *dependent variable* is the age at which the child used the toilet at least 90% of the time.

   Results: Both learned at about the same age, i.e., 850 days.

   Conclusion: Toilet training before about 650 or 700 days of age has no benefit.

3.   *Can I state the theoretical explanation?*

   We are able to learn certain skills only when genetically determined maturation has taken place. Based on the instructions for maturation in the child's genes, ability to control one's bowels and bladder is not available until about 650 or 700 days from birth.

Table 5.4. Evaluating Maturation

1. *Can I evaluate the assertion?*
    Clarity: Maturation is a circular concept, which can be used to "explain" almost
    any failure to learn. The definition will remain unmeasurable until scientists
    discover the genetic code for maturation.
    Testability: For any newly acquired behavior, it is difficult to separate the
    contributions of experience (including training) and maturation.
    Importance: The assertion is very important for parents and teachers who want to
    promote successful learning in children.

2. *Can I evaluate the empirical evidence?*
    Name: McGraw (1940)
    Method: There were only two subjects. There is possible experimenter bias since
    the parents may treat the boys differently, or may use a different definition
    of "successful training" for each boy.
    Results: No detail is given about what "success in percent" means. Why is Hilton's
    progress more dramatic and consistent? It looks like Hugh began to make
    progress between 600 and 650 days, not 700 as reported in the text.
    Conclusion: What about the long-term performance of the boys? Perhaps Hilton
    will show some advantages over Hugh as they get older.

3. *Can I evaluate the theoretical explanation?*
    On the surface, the results seem to be a strong confirmation of the role of
    maturation in learning. However, an alternative explanation is that the parents
    learned how to be effective toilet trainers while training Hugh so they were more
    efficient in training Hilton. Another alternative explanation is that the parents had
    different expectations for the two boys, i.e., they did not expect a one-year-old to
    learn and so did not try as hard.

Cognitive Psychology

Do You Understand the Argument?

Now, you are ready to apply the strategies you have learned to real passages from introductory psychology textbooks. Please carefully read the passage on *Short-Term Memory* that is presented in box 6.1. Then, turn to table 6.1 and fill in the spaces, indicating how you summarize the assertion, evidence, and explanation given in the passage. If you have difficulty, you should reexamine table 3.1.

▓▓▓▓   **Box 6.1.** Short-Term Memory

When a bit of information is selected for further processing, it's transferred into *short-term memory,* or *STM.* It is not necessary to intentionally transfer information to STM; generally, just paying attention to the information is enough to transfer it. You may not intentionally try to memorize the price of your dinner, but you will recognize that you were given the wrong change even though you did not try to memorize the price of the dinner.

Rehearsal in Short-Term Memory
As the name implies, STM is good for only temporary storage of information. In general, information fades from STM in less than half a minute unless it is "renewed," and it often fades in only a few seconds (Ellis & Hunt, 1989). Fortunately, information can be renewed in STM by mental repetition, or *rehearsal,* of the information. When a grocery list is rehearsed regularly in this way, it can be held in STM for relatively long periods of time. If the list is not rehearsed, however, it is soon lost. Rehearsing the information stored in STM has been compared to juggling eggs: The eggs stay in perfect condition as long as you keep juggling them, but as soon as you stop juggling, they are lost.

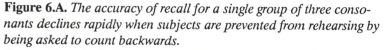

Figure 6.A. *The accuracy of recall for a single group of three consonants declines rapidly when subjects are prevented from rehearsing by
being asked to count backwards.*
*Source: From L. R. Peterson and M. J. Peterson, "Short-Term Retention
of Individual Items" in* Journal of Experimental Psychology, 58, 193-
198, 1959.

Our first reliable estimate of the limited life span of information in STM was
provided by an experiment conducted by Lloyd and Margaret Peterson (1959). The
subjects were shown a single combination of three consonants (such as LRP) and asked
to remember it as they counted backward by threes to keep them from rehearsing the
letters. The subjects counted backward for brief intervals (0 through 18 seconds) and
then were asked to recall the letters. As shown in figure 6.A, the subjects were able to
remember the three consonants less than 20 percent of the time after only 12 seconds
had passed. These findings make it clear that memories are quite impermanent in STM
unless kept alive by rehearsal.

Table 6.1. Do I Understand the Argument?

1a.   Can I state the assertion (and underline the key terms)?

1b.   Can I put the assertion in my own words?

1c.   Can I give a concrete example of the assertion (including underlining and labeling examples of the key terms)?

2.   Can I state the empirical evidence?
Name:

Method (including underlining and labeling the independent and dependent
            variables):

Results:

Conclusion:

3.   Can I state the theoretical explanation?

Table 6.2. Can I Evaluate the Argument?

1. *Can I evaluate the assertion?*
    Clarity:

    Testability:

    Importance:

2. *Can I evaluate the empirical evidence?*
    Name:

    Method:

    Results:

    Conclusion:

3. *Can I evaluate the theoretical explanation?*

Can You Evaluate the Argument?

Next, fill in the spaces in table 6.2, indicating how you criticize the assertion, evidence, and explanation given in the text. If you have difficulty, you should reexamine table 3.5.

How Well Did You Do?

Were you able to understand and evaluate the argument presented in the *Short-Term Memory* passage? Take a moment to compare your summaries with those given in table 6.3 and table 6.4. For each number 1 through 3 in table 6.3 and table 6.4, write "yes" if your answer was better or equivalent to our answer and write "no" if your answer was worse than our answer. Summarize your scoring by writing "yes" or "no" for each of the following questions:

___ Did I understand the assertion?
___ Did I understand the evidence?
___ Did I understand the explanation?
___ Could I evaluate the assertion?
___ Could I evaluate the evidence?
___ Could I evaluate the explanation?

If you answered "no" for any question, use the following space to write down new points you would like to include.

Table 6.3. Understanding Short-Term Memory

1a.   *Can I state the assertion (and underline the key terms)?*

Short-term memory has the property of being able to hold pieces of information for only a *brief* period without *rehearsal.*

1b.   *Can I put the assertion in my own words?*

I can remember new pieces of information for no more than about 20 seconds unless I actively keep reciting each piece of information.

1c.   *Can I give a concrete example of the assertion (including labeling examples of key terms)?*

I go to a new town and look up my friend's telephone number in the phone book at a pay phone; I insert the coin and dial the number. The line is busy, and, in the *time it has taken me to dial and listen* (brief period), I have forgotten the *telephone number* (information in short-term memory).

2.   *Can I state the empirical evidence?*

Name: Peterson & Peterson (1959)

Method: Subjects were shown three letters and asked to recall them 0 to 18 seconds later. During the interval, the subjects had to count backward by threes from a given number, in order to prevent rehearsal. The *independent variable* is the number of seconds in the retention interval; the *dependent variable* is the percent correctly recalled.

Results: Recall accuracy fell from nearly perfect at 0 seconds to nearly zero by 18 seconds.

Conclusion: Unrehearsed information fades from short-term memory within 18 seconds.

3.   *Can I state the theoretical explanation?*

Traces that register the information in short-term memory fade within a few seconds. The traces can be strengthened by paying attention to them.

Table 6.4. Evaluating Short-Term Memory

1.   *Can I evaluate the assertion?*

    Clarity: How do we know that we have or have not prevented rehearsal? How do we know what constitutes a piece of information in short-term memory?

    Testability: The mechanism underlying the limits on short-term memory is not clear, so it is hard to test it.

    Importance: Research on how humans process information is important, and short-term memory seems to be at the center of human information processing.

2.   *Can I evaluate the empirical evidence?*

    Name: Peterson & Peterson (1959)

    Method: Counting backward by threes could be distracting to the subject; we need an intervening task that prevents rehearsal but does not cause interference. The material used is not representative of real memory situations.

    Results: A careful analysis of the data (not shown in the textbook) indicates that subjects perform better on the first few items during the session than the last few items; this cannot be accounted for using the concept of fading in short-term memory.

    Conclusion: The results show a consistent decrease in recall performance as a function of retention interval. However, how does this conclusion square with the fact that people can remember information for a very long time if they try to?

3.   *Can I evaluate the theoretical explanation?*

    The results may be accounted for by saying that counting backward by threes filled the subjects' short-term memories with numbers that displaced the original material; therefore, material was forgotten because it was displaced rather than because it automatically faded.

Motivation

Do You Understand the Argument?

Now, you are ready to apply the strategies you have learned to real passages from introductory psychology textbooks. Please carefully read the passage on the *Yerkes-Dodson Law* that is presented in box 7.1. Then, turn to table 7.1 and fill in the spaces, indicating how you summarize the assertion, evidence, and explanation given in the passage. If you have difficulty, you should reexamine table 3.1.

▒▒▒ **Box 7.1.** Yerkes-Dodson Law

The Nature of Arousal

As you drive down a highway, the fog thickens. Suddenly you see a pile of cars in front of you. Your mind temporarily freezes, your muscles tighten, your stomach becomes queasy, and your heart feels like it is going to pound out of your chest. You immediately slam on the brakes and try to veer away from the pile of cars. Tires screech, windshield glass flies, and metal smashes. Then all is quiet. After a few short seconds you realize you are alive. You find that you can walk out of the car. Your fear turns to joy as you sense your luck in not being hurt. In a couple of seconds, the joy turns to anger. You loudly ask who caused the accident.

As you moved through the emotions of fear, joy, and anger, your body changed. During intense arousal your sympathetic nervous system was working. At the time of the accident your arousal decreased as the parasympathetic nervous system became more dominant: your heart rate, breathing rate, and blood sugar level decreased; your pupils constricted; and your stomach secretion and your salivation increased. The sympathetic and parasympathetic nervous systems are the two divisions of the autonomic nervous system.

Early in this century two psychologists described the role of arousal in performance. What is now known as the Yerkes-Dodson Law states that performance is best under conditions of moderate rather than low or high arousal. At the low end of arousal you might be too lethargic to perform tasks well; at the high end you may not be able to concentrate. Think about how aroused you were the last time you took a test. If your arousal was too high, your performance probably suffered.

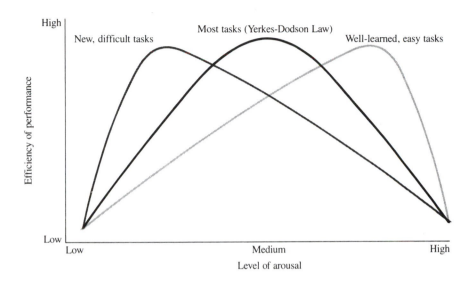

Figure 7.A. *Arousal and performance. The Yerkes-Dodson Law states that optimal performance occurs under moderate arousal. However, for new or difficult tasks, low arousal may be best, for well-learned, easy tasks, high arousal can facilitate performance.*

Moderate arousal often serves us best in tackling life's tasks, but there are times when low or high arousal produces optimal performance. For well-learned or simple tasks (signing your name, pushing a button on request), optimal arousal can be quite high. By contrast, when learning a task (such as how to play tennis) or doing something complex (such as solving an algebraic equation), much lower arousal is preferred. Figure 7.A projects how arousal might influence easy, moderate, and difficult tasks. As tasks become more difficult, the ability to be alert and attentive, but relaxed, is critical to optimal performance.

Table 7.1. Do I Understand the Argument?

1a.   *Can I state the assertion (and underline the key terms)?*

1b.   *Can I put the assertion in my own words?*

1c.   *Can I give a concrete example of the assertion (including underlining and labeling examples of the key terms)?*

2.   *Can I state the empirical evidence?*
        Name:

        Method (including underlining and labeling the independent and dependent
               variables):

        Results:

        Conclusion:

3.   *Can I state the theoretical explanation?*

Table 7.2. Can I Evaluate the Argument?

1.   *Can I evaluate the assertion?*
     Clarity:

     Testability:

     Importance:

2.   *Can I evaluate the empirical evidence?*
     Name:

     Method:

     Results:

     Conclusion:

3.   *Can I evaluate the theoretical explanation?*

Can You Evaluate the Argument?

Next, fill in the spaces in table 7.2, indicating how you criticize the assertion, evidence, and explanation given in the text. If you have difficulty, you should reexamine table 3.5.

How Well Did You Do?

Were you able to understand and evaluate the argument presented in the *Yerkes-Dodson Law* passage? Take a moment to compare your summaries with those given in table 7.3 and table 7.4. For each number 1 through 3 in table 7.3 and table 7.4, write "yes" if your answer was better or equivalent to our answer and write "no" if your answer was worse than our answer. Summarize your scoring by writing "yes" or "no" for each of the following questions:

____ Did I understand the assertion?
____ Did I understand the evidence?
____ Did I understand the explanation?
____ Could I evaluate the assertion?
____ Could I evaluate the evidence?
____ Could I evaluate the explanation?

If you answered "no" for any question, use the following space to write down new points you would like to include.

Table 7.3. Understanding the Yerkes-Dodson Law

1a.   *Can I state the assertion (and underline the key terms)?*

        A person's *level of arousal* is related to his *efficiency of performance*: for most tasks, performance is best under medium levels of arousal; for easy tasks, higher levels of arousal are best; for hard tasks, lower levels of arousal are best.

1b.   *Can I put the assertion in my own words?*

        I perform most tasks best when I am neither too relaxed nor too stressed; but for easy, well-learned tasks I perform best when I am under some pressure, and for hard tasks I perform best when I am under no pressure.

1c.   *Can I give a concrete example of the assertion (including labeling examples of key terms)?*

        A *star baseball player* (easy, well-learned task) is most likely to *get a hit* (efficient performance) when he comes up to bat in the *bottom of the ninth inning with bases loaded, two outs, and the winning run on base* (high level of arousal).

2.   *Can I state the empirical evidence?*

Name: No specific reference is given.

Method: No specific experiment is described; however, apparently the *independent variables* are level of arousal and difficulty of task whereas the *dependent variable* is efficiency of performance.

Results: Presumably, the results of experiments would be consistent with the figure in the text and, thus, with the description of the law given previously.

Conclusion: Presumably, the conclusion is that the Yerkes-Dodson law is correct; however, only anecdotal evidence is given.

3.   *Can I state the theoretical explanation?*

        Arousal is linked to the activity of the autonomous nervous system, but the text does not give an explanation of how this system works.

Table 7.4. Evaluating the Yerkes-Dodson Law

1. *Can I evaluate the assertion?*

   Clarity: The term "arousal" is vague: how do we know when it is "low," "medium," or "high"? The measurement of performance is not well-defined: when is performance "low" or "high"? The level of difficulty of tasks is not defined: what makes a task "easy" or "hard"?

   Testability: If the results of a study produce an inverted U, we can say the theory is supported. If the results produce a rising line or a falling line, we can say that we have not examined a wide enough range of arousal. Thus, it may not be possible to find data that refute the law.

   Importance: The relation between arousal and performance is important for education and coaching.

2. *Can I evaluate the empirical evidence?*

   Name: Yerkes-Dodson (No specific reference is given.)

   Method: None is given.

   Results: Only anecdotes are given. The graph, apparently, is idealized.

   Conclusion: The conclusion makes intuitive sense but cannot be evaluated based on material presented in the textbook.

3. *Can I evaluate the theoretical explanation?*

   The explanation is not supported by any solid evidence, nor is it refuted. The physiological mechanisms underlying the law are not clearly presented so there is really no theoretical mechanism to evaluate.

Biopsychology

Do You Understand the Argument?

Now, you are ready to apply the strategies you have learned to real passages from introductory psychology textbooks. Please carefully read the passage on pain that is presented in box 8.1. Then, turn to table 8.1 and fill in the spaces, indicating how you summarize the assertion, evidence, and explanation given in the passage. If you have difficulty, you should reexamine table 3.1.

▓▓▓▓ **Box 8.1   Pain**

An injury or intense stimulation of sensory receptors induces pain. So, bright lights, loud noises, hot spices, and excessive pressure, as well as cuts, burns, and bruises, are painful.

Many pain receptor neurons transmit pain impulses by releasing the neurotransmitter *substance P* from their axons. For example, the intensity of arthritis pain varies with the amount of substance P released by neurons that convey pain impulses from the joints (Levine et al., 1984). People with a rare medical disorder that reduces their sensitivity to pain have unusually low levels of substance P at synapses that transmit neural impulses from pain receptors (Pearson, Brandeis, & Cuello, 1982).

The most influential theory of pain is the gate-control theory formulated by psychologist Ronald Melzack and biologist Patrick Wall (1965). The theory assumes that pain impulses from the limbs or body pass through a part of the spinal cord called the *substantia gelantinosa*, which provides a "gate" for pain impulses. Stimulation of neurons that convey touch sensations "closes" the gate, preventing input from neurons that convey pain sensations. This might explain why rubbing a shin that you have banged against a table will relieve the pain. The closing of the pain gate is stimulated by the secretion of *endorphins*, which are the brain's natural opiates. Endorphins might close the gate by inhibiting the secretion of substance P (Ruda, 1982).

The pain gate is also affected by neural impulses that originate in the brain (Dubner & Bennett, 1983). This might explain why anxiety, relaxation, and other psychological factors can affect pain perception. For example, it may explain the so-called Anzio effect, in which wounded soldiers returning from the fierce World War II battle for control of Anzio, Italy, needed less morphine than did civilians with similar wounds. Perhaps because the soldiers interpreted their wounds as tickets away from the battlefield, they experienced their pain as less intense (Wallis, 1984). Their pain might have been reduced by neural impulses sent from the brain to the substantia gelatinoisa, closing the pain gate. The Anzio effect also shows that the *reinterpretation* of pain can reduce its intensity (Devine & Spanos, 1990).

Table 8.1. Do I Understand the Argument?

1a.   *Can I state the assertion (and underline the key terms)?*

1b.   *Can I put the assertion in my own words?*

1c.   *Can I give a concrete example of the assertion (including underlining and labeling examples of the key terms)?*

2.   *Can I state the empirical evidence?*
       Name:

       Method (including underlining and labeling the independent and dependent variables):

       Results:

       Conclusion:

3.   *Can I state the theoretical explanation?*

Can You Evaluate the Argument?

Next, turn to table 8.2 and fill in the spaces, indicating how you criticize the assertion, evidence, and explanation given in the text. If you have difficulty, you should reexamine table 3.5.

How Well Did You Do?

Were you able to understand and evaluate the argument presented in the pain passage? Take a moment to compare your summaries with those given in table 8.3 and table 8.4. For each number 1 through 3 in table 8.3 and table 8.4, write "yes" if your answer was better or equivalent to our answer and write "no" if your answer was worse than our answer. Summarize your scoring by writing "yes" or "no" for each of the following questions:

___ Did I understand the assertion?
___ Did I understand the evidence?
___ Did I understand the explanation?
___ Could I evaluate the assertion?
___ Could I evaluate the evidence?
___ Could I evaluate the explanation?

If you answered "no" for any question, use the following space to write down new points you would like to include.

Table 8.2. Can I Evaluate the Argument?

---

1.   *Can I evaluate the assertion?*
      Clarity:

      Testability:

      Importance:

2.   *Can I evaluate the empirical evidence?*
      Name:

      Method:

      Results:

      Conclusion:

3.   *Can I evaluate the theoretical explanation?*

65

Table 8.3. Understanding Pain

1a.   *Can I state the assertion (and underline the key terms)?*
       Endorphins released by neurons n the brain may stimulate other neurons to send signals that close the *pain gate* in the spinal cord, preventing input from neurons than convey pain sensations.

1b.   *Can I put the assertion in my own words?*
       Incoming pain signals from sense receptors will not be experienced as strongly if the pain gate is closed by signals from the brain.

1c.   *Can I give a concrete example of the assertion (including labeling examples of key terms)?*
       Rubbins a shin that you have just banged against a table will relieve the *pain,* because this stimulation closes the *gate.*

2.   *Can I state the empirical evidence?*
Name: Wallis (1984)
Method: Researchers recorded the amount of morphine needed to relieve the pain (dependent variable) of soldiers and civilians with similar battle wounds (independent variable).
Results: Soldiers required less morphine, and thus experienced lower levels of pain, than civilians with similar wounds.
Conclusion: Soldiers interpreted their wounds as tickets away from the battlefield, so their brains sent messages that closed the pain gate. Psycological factors affect the experience of pain by closing or opening the pain gate.

3.   *Can I state the theoretical explanation?*
       The emotional excitement of battle signals the pain gate to close. Soldiers in battle may produce a drug that blocks the experience of pain.

Table 8.4. Evaluating Pain

---

1.   Can I evaluate the assertion?

Clarity: The definition of pain is unclear: how is the level of pain measured? The
amount of morphine required is not a perfect measure of pain because some
people are more sensitive to morphine than others. The conditions for opening
and closing the pain gate are unclear: how can we tell in advance whether a
person will experience an emotional state that closes the pain gate?

Testability: There are ethical problems in doing controlled research on pain. Most
of the situations described in the textbook are hypothetical or anecdotal; it
would be hard to recreate them within a research context.

Importance: This line of research is important for the treatment of pain.

2.   Can I evaluate the empirical evidence?

Name: Wallis (1984)

Method: The participants were not randomly assigned to groups so possibly the
soldiers were less sensitive to pain than the civilians. Perhaps the doctors
tend to use less morphine on soldiers than civilians. Perhaps the wounds were
really not equivalent in soldiers and civilians. Perhpas the soldiers felt as
much pain as the civilians but were less likely to complain about the pain.
Method is not reported in enough detail, e.g., how many people were in the
study?

Results: Results are not reported in enough detail, how much less morphine did the
soldiers need as compared to the civilians?

Conclusion: The conclusion seems tenuous because there is not much solid evidence.
It is assumed that morphine somehow helps to close the pain gate.

3.   Can I evaluate the theoretical explanation?

The physiological mechanisms is not clearly specified. Some alternative
explanations are: (a) If the soldiers are in shock, they may not feel pain. (b) If
the soldiers want to look tough, they may not admit feeling pain. (c) Rather than
stopping pain signals at a pain gate, we may simply ignore pain messages once they
reach the brain.

Although the tables 8.3 and 8.4 examine one important argument
concerning the gate-control theory, there are several other arguments
presented in the lesson. For example, another assertion is that: pain
receptors transmit pain impulses by releasing the neurotransmitter
substance P; the supporting evidence is that the experience of pain of
arthritis varies with the amount of substance P released from the joints;
and the theoretical explanation is that substance P is interpreted at the
brain as pain.

Glossary

argument An assertion with supporting empirical evidence and a supporting theoretical explanation.

assertion A general description of the characteristics of one or more things or the relation between two or more things. Four types of assertions are relational, causal, property, and differential. See *argument*.

causal assertion A type of assertion of the form "A causes B." See *assertion*.

critical thinking An active and systematic attempt to understand and evaluate arguments.

critical thinker A person who engages in critical thinking, including the use of strategies for understanding and evaluating arguments.

descriptive evidence A type of empirical evidence that is based on making observations of naturally occurring events. See *empirical evidence*.

differential assertion A type of assertion of the form "A differs from B on property P." See *assertion*.

empirical evidence Specific observations that support or refute an assertion. Four types of empirical evidence are quantitative-descriptive, qualitative-descriptive, quantitative-experimental, and qualitative-experimental. See *argument*.

evaluating an argument Being able to judge the clarity, testability, and importance of an assertion; identifying flaws in the method, results, or conclusion of empirical evidence; and generating alternative theoretical explanations.

evidence See *empirical evidence*.

experimental evidence A type of empirical evidence based on observations from the results of an experiment. See *empirical evidence*.

explanation See *theoretical explanation*.

learning strategies Activities engaged in by the learner for the purpose of improving the process of learning.

property assertion A type of assertion of the form "P is a property of A." See *assertion*.

qualitative evidence A type of empirical evidence expressed in words or pictures. See *empirical evidence*.

quantitative evidence A type of empirical evidence expressed in numbers. See *empirical evidence*.

relational assertion A type of assertion of the form "A and B are related." See *assertion*.

strategies See *learning strategies, thinking strategies*. See *evaluating an argument, understanding an argument*.

theoretical explanation A hypothesized mechanism or principle that logically justifies or refutes an assertion. See *argument*.

thinking strategies Behaviors engaged in by the learner for the purpose of improving the process of thinking.

understanding an argument Being able to summarize an assertion, identify supporting evidence, and cite a theoretical explanation.

Bibliography

Collins, C. & Mangieri, J.N. (1992). *Teaching thinking: An agenda for the 21st century.* Hillsdale, NJ: Erlbaum.

Cook, L.B. & Mayer, R.E. (1988). Teaching readers about the structure of scientific text. *Journal of Educational Psychology, 80* 448-456.

Ennis, R. H. (1962). A concept of critical thinking. *Harvard Educational Review, 32,* 81-111.

Fleet, J., Goodchild, F., & Zajchowski, R. (1989). *Learning for success: Skills and strategies for Canadian students.* Toronto: Harcourt-Brace Jovanovich.[1]

Halpern, D.F. (1989). *Thought and kowledge: An introduction to critical thinking (2nd ed.).* Hillsdale, NJ: Erlbaum.

Kintsch, W. (1986). Learning from text. *Cognition and Instruction, 3,* 87-108.

Lochhead, J. (1979). An introduction to cognitive process instruction. In J. Lochhead & J. Clement (Eds.), *Cognitive process instruction: Research on teaching thinking skills.* Philadelphia: Franklin Institute Press.

Mayer, R. E. (1987). *Educational psychology: A cognitive approach.* Boston: Little, Brown.

Mayer, R. E. (1989). Teaching for thinking: Research on the teachability of thinking skills. In *G. Stanley Hall Lecture Series,* Vol. 9. Washington, DC: American Psychological Association.

McPeck, J. (1981). *Critical thinking in education.* New York: St. Martin's Press.

Meyer, B. J. F. (1991), Basic research on prose comprehension: A critical review. In D.F. Fisher & C.W. Peters (Eds.), *Comprehension and the competent reader* (pp. 8-35). New York: Praeger.

Norman, D. A. (1980). Cognitive engineering and education. In D. T. Tuma & F. Reif (Eds.), *Problem solving and education: Issues in teaching and research.* Hillsdale, NJ: Erlbaum.

Northey, M. & Timney, B. (1986). *Making sense in psychology and the life sciences.* Toronto: Oxford University Press.

Perkins, D., Jay, E., & Tishman, S. (Eds.). (1993). Special issue: New conceptions in thinking. *Educational Psychologist, 28,* 1-85.

Pressley, M. (1990). *Cognitive strategy instruction that really improves childen's academic performance.* Cambridge, MA: Brookline Books.

Stice, J. E. (1987). *Developing critical thinking and problem-solving abilities.* San Francisco: Josey-Bass.

Weinstein, C. E. & Mayer, R. E. (1985). The teaching of learning strategies. In M. C. Wittrock (Ed.), *Handbook of research on teaching* (3d ed.). New York: Macmillan.

Young, R. E. (1980). *Fostering critical thinking.* San Francisco: Josey-Bass.

---

[1]Fleet, Goodchild, & Zajchowski (1989) present an argument-based framework for use in a wide variety of subject matter domains.

Visual Credits